"Why do engineers love GLP FE/EIT study guides?"

Reader remarks from the GLP in-basket....

"I greatly appreciate your coming out with such an excellent *FE/EIT Discipline Review*. I walked into the exam feeling prepared and confident...and most important, I passed!"
—Bill Johnston, Raleigh, NC

"After looking at several other books, I am convinced your *Fundamentals of Engineering* is the best review around. It is very well organized, thorough, and concise. An outstanding book."
—Paul Griesmer, Cleveland, OH

"This is a time- and cost-effective tool. I easily passed the FE on my first attempt."
—Omar Mureebe, Uniroyal Chemical Company, Middlebury, CT

"In your FE book you cut to the chase and don't embellish it... Yours is the only review that doesn't scare the pants off people. It's exactly what I want for teaching my review course."
—John Thorington

"I was talking to a student a few hours ago and he said the morning problems on the FE/EIT were like the sample problems in your book—verbatim. It's the best review I've ever seen. When we started using your review in 1994, our students' pass rates went up 20%! Now we will continue to use your book because we take professional licensing seriously."
—Dr. Gary Rogers, Virginia Military Institute

"Your FE study guide is very thorough, well-organized and easy to use. It was instrumental in helping me brush up on engineering material I had not seen for 5 years. I passed the FE the first time!"
—Brian Campbell, Project Engr, Trailmaster Corp, Ft. Worth, TX

"Thank you for the book! I think I passed the test because of it. Stuff I hadn't studied for more than 15 years came back to me clearly. The two practice exams are what helped me the most."
—Juan Alfaro, College Park, MD

"Three years out of school and all I needed to pass the FE exam, on the first try, was your review book! I have recommended it to everyone I know. I look forward to using your PE reference."
—David Robins, Syracuse, NY

"Your book is great! Very easy to use. Although I only had time to skim through it, it helped me review the very basics I needed to know (i.e., Chemistry, Econ). I passed without a problem! Thanks! I'm keeping it as a handy reference."
—Julie Volby, Civil Engineering Student, Univ of Minnesota, Minneapolis, MN

"Your review material is perfect for the non-traditional 'night student.' It refreshes and restores many years of part-time study, without being overly exhaustive."
—Jerome Bobak, Designer Mechanical Engineer, Grand Island, NY

"This book provided all the necessary information for passing the FE exam. I was able to review for and *pass* the exam in only 3 weeks!"
—Joseph Rozza, Recent Graduate, Orlando, FL

Important Information

FE examination date: _____ time: _____

location: _____

_____

Examination Board Address: _____

_____

_____

phone: _____

Application was requested on this date: _____

Application was received on this date: _____

Application was accepted on this date: _____

This book belongs to: _____

_____

_____

_____

phone: _____

·FE/EIT·
Electrical Engineering Review

5th edition

Merle C. Potter, Phd, PE—Editor

An Efficient Review for the Afternoon Test in Electrical Engineering

From the professors who know it best...

H. Roland Zapp	*Electromagnetic Fields*
Dennis Wiitanen	*Electrical Power*
H. Roland Zapp	*Communications and Signal Processing*
Bong Ho	*Solid State Devices*
R. Lal Tummala	*Computer Engineering*

Authors are professors at Michigan State University, with the exception of Wiitanen who teaches at Michigan Tech.

published by:

GREAT LAKES PRESS

Okemos, Michigan   Wildwood, Missouri
P.O. Box 550, Wildwood MO 63040
Customer Service (636) 273-6016   www.glpbooks.com

International Standard Book Number 1-881018-38-5

Copyright © 2001 by Great Lakes Press, Inc.

All rights are reserved.
No part of this publication may be reproduced in any form or
by any means without prior written permission of the publisher.

All comments and inquiries should be addressed to:
   Great Lakes Press
   PO Box 550
   Wildwood, MO 63040-0550
   Phone (636) 273-6016
   Fax (636) 273-6086
   www.glpbooks.com
   custserv@glpbooks.com

Library of Congress Control Number: 2001091932

Printed in the USA by Sheridan Books, Inc. of Ann Arbor, Michigan.

10 9 8 7 6 5 4 3 2 1

Table of Contents

A Brief Outline of This Review 3
State Boards of Registration Information 10

Passing the FE/EIT Exam

1. Electromagnetic Fields .... Zapp 17

2. Electric Power .... Wiitanen 23
   2.1 Balanced Three Phase Systems 23
   2.2 AC Machines 26

3. Communications and Signal .... Zapp 29

4. Solid State .... Ho 35

5. Computer .... Tummala 43
   5.1 Number Systems and Codes 43
   5.2 Binary Addition and Subtraction 46
   5.3 Logic Operations and Boolean Algebra 48
   5.4 Karnaugh Maps 51
   5.5 Flip-Flops 53

6. Control Systems .... Tummala 57
   6.1 Transfer Functions 57
   6.2 Poles, Zeros and Stability 60
   6.3 Time Domain Performance 62
   6.4 Steady-State Response 65
   6.5 Root Locus 67
   6.6 Frequency Response 72

The FE/EIT Review for Electrical Engineering

Electrical Engineering Discipline Exam 81

EE Practice Exam

Appendix A—NCEES Equation Summaries 113
Appendix B—Units and Conversions 125

Appendixes: Equations & Units

Preface

by Merle C. Potter

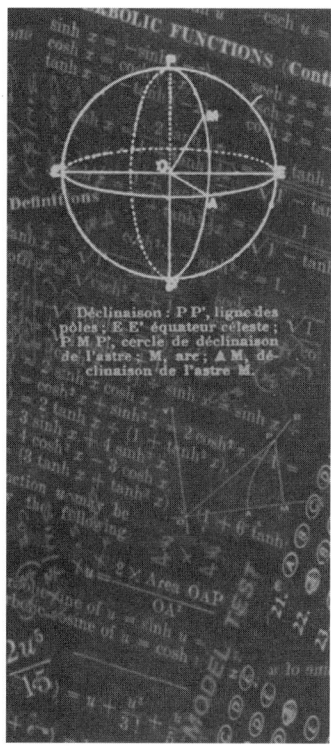

The FE/EIT Exam Format

The FE/EIT exam is a basic competency exam that consists of a General engineering test during the morning session, covering twelve subject areas required in most engineering curricula, and an afternoon session that is discipline-specific. The afternoon session will consist of one of six discipline-specific tests: Chemical, Civil, Electrical, Industrial, Mechanical, or General. The General test is intended for those engineers who do not fall into any of the five specific disciplines. Each examinee will take the morning General test and one of the afternoon tests.

Since there are over 30 engineering disciplines, examinees in over 25 of those disciplines will be expected to take the General test during the afternoon session. Most of the state boards will recognize a "pass" on any of the afternoon tests by any examinee. Consequently, it may be advisable for you to elect the afternoon General test even though you are a Civil, Electrical, Mechanical, Chemical, or Industrial engineer. Check with your state board if you are uncertain of your state's position.

Review Overview

This book was written as a review for the afternoon test of the FE/EIT exam in Electrical engineering. Our strategy is to offer a short review and illustrate each subject with at least one example problem which is as exam-like as possible; this has been done for the subjects and primary equations included in the official NCEES Reference Handbook. Since the Handbook is the only book allowed into the examination, this review should cover nearly all problems tested on the afternoon session of the FE/EIT exam.

We offer a full 60-question practice test, with full solutions. This exam should help you decide if you should take the discipline test in your area or select the General test option.

NCEES Handbook

Many of the afternoon questions can be answered by referring to equations, tables, or charts included in the first 83 pages of the NCEES Reference Handbook —material that is primarily intended for the General tests. We recommend that you use our *Fundamentals of Engineering Review*, a book written to help you prepare for the morning General test, and much of the afternoon discipline tests, as well as the afternoon General test. Call 1-800-837-0201 to order a copy. We also offer HP 48GX preprogrammed calculators and a companion 240-page *Jump Start the HP 48G/X* helpbook—highly effective aids to help you pass the FE/EIT exam.

Free CD Tests!

Finally, to further help you decide if you are ready to take the test in your discipline, we have included a coupon at the back cover of this book for a free CD which presents CE, ME, EE and IE practice exams in a user-friendly, rather attractive interactive environment. The CD will interpret the results of your test, suggesting if you are ready for the actual exam and indicating the score you obtained in each of the subject areas of the exam. It also gives you study strategy suggestions.

To obtain your copy of the free CD exams, simply fill out the coupon at the back of the book and drop it in the mail.

Best wishes in your study and on your test day!

Dr. Merle C. Potter, PE
Okemos, Michigan

Passing the FE/EIT Exam

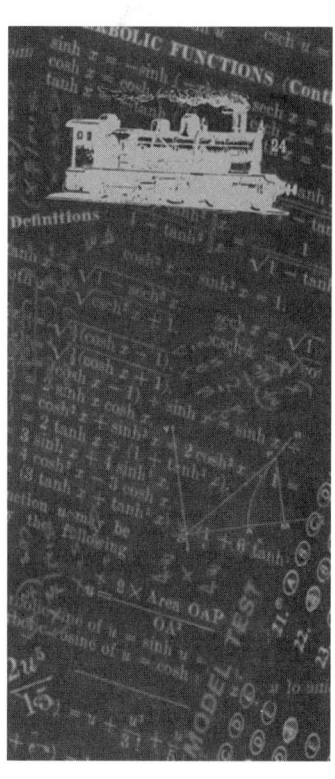

This book presents an efficient review of the major Electrical Engineering subjects tested in the afternoon session of the FE/EIT exam. GLP offers a companion book, *Fundamentals of Engineering Review,* which presents a succinct yet complete review of the 12 major topics of the morning General session, which you'll also have to take. You may order a copy of our General review by calling 1-800-837-0201.

A Brief Outline of This Review

This book contains:
- Short, succinct reviews of the important aspects of most of the Electrical Engineering subject areas—just enough coverage to ensure that all-important pass! Some subject areas are presented with sample problems only.
- Sample problems that present the situational use of the more important equations presented in the NCEES Handbook.
- Complete solutions to each sample problem showing the details needed to arrive at a solution.
- A full 60-question practice test, with solutions.
- Mail-in coupon for *free* CD containing exams for 4 different disciplines, plus our unique study strategy software called Study-Director™.

In 1983, we at Great Lakes Press developed an FE/EIT review with the cooperation of a team of colleagues in response to our experience in coordinating FE/EIT review courses. The kind of well-planned material that an engineer naturally desires to use did not exist at the time. The options were either to pick and

Why We Created This FE Review

choose from a large, encyclopedia-like book that covers almost every topic in engineering, or to use material that didn't even attempt to cover the topics tested on the exam. It was either feast or famine, and it was all awkward—and often rather expensive! So we recruited popular lecturers from the university campus and prepared our first *Fundamentals of Engineering Review*.

With the announcement of the new exam format in 1996, we quickly pulled together a team of 20 professors to prepare discipline-specific reviews in time for the very first test in the new format.

On a daily basis we act as advocate for the test-taker in our involvement with engineering departments, associations, review courses and registration governing bodies. We are dedicated to making the entire licensing and registration process as reasonable as possible. We've responded faster than any other reviewing option to all of the many exam changes that keep coming at test-takers.

Our strategy for beating the Discipline-Specific format is a prime example of our leadership. We've prepared this review focusing solely on the equations, figures, and tables presented in the NCEES Reference Handbook. This seems to be the most reasonable way to prepare for the very many topics covered by discipline-specific tests. It is vital for effective familiarization not to get too detailed in one's study for such broad exams. Also, the Handbook is the only aid allowed into the exam besides a calculator. There will be questions that may not require the use of the equations from the Handbook; these will be based on general information from the respective courses. Such general information may not be included in this review.

Our "Pass or Your Money Back" Guarantee!

For your added confidence, with GLP you're guaranteed to pass! If you use our preps and fail the FE exam, send us your name & address, receipt, book & CD, and copy of failure notice (within 30 days of notice)—and we will issue you a full refund.

How to Become a Professional Engineer

To become registered as an engineer, a state may require that you:
1. Graduate from an ABET-accredited engineering program
2. Pass the *Fundamentals of Engineering* exam
3. Pass the *Principles and Practice of Engineering* (PE) exam after several years of engineering experience

Requirements vary from state to state, so you should obtain local guidelines and follow them carefully.

Why Get an Engineering License?

Registration is necessary if an engineer works as a consultant, and is highly recommended in certain industries—especially when one is, or hopes to be, in a management position.

How the FE Exam Is Scored

In the morning session, each of the 120 problems is worth <u>one-half</u> point. Thus a maximum score of 60 is possible. The afternoon session is different from the morning in that each problem is worth <u>one</u> point. With half the number of questions, the maximum possible score in the afternoon session is also 60.

The two-session total is 120 points, with both sessions having equal weight. A predetermined passing percentage is not established, nor is the exam graded on a curve. From recent history, a score of 60 (50% correct) would probably be a passing score. Each year the exact score is determined by statistical methods, so the score of 60 is an approximation.

Examination Format

The FE is composed of a four-hour general-subject morning session, followed by a one-hour break, then a four-hour discipline-specific afternoon session of 60 questions. The two sessions have a total of 180 4-part multiple-choice questions covering the subject areas listed below.

Morning Session

Subject Area	Approximate Number
Chemistry	11
Computers	6
Dynamics	10
Economics	5
Electrical Circuits	12
Ethics	5
Fluids	8
Materials Science	8
Mathematics	24
Mechanics of Materials (Solids)	8
Statics	12
Thermodynamics	11

Total Questions: 120

Subject Lists of the Discipline-Specific Afternoon Tests

General • Chemistry, Computers, Dynamics, Economics, Electrical, Ethics, Fluids, Material Science, Mathematics, Mechanics of Materials, Statics, Thermo

Civil • Construction, Enviro, Hydraulics, Hydrology, Numerical Methods, Legal, Soils, Foundations, Structural, Design, Surveying, Water Treatment, Waste Water

Mechanical • Controls, Computers, Systems, Energy Conversion, Power Plants, Turbomachines, Fluids, Heat Transfer, Materials, Instrumentation, Design, HVAC, Solids, Thermodynamics

Electrical • Analog, Communication, Numerical Methods, Computer Hardware, Software, Controls, Systems, Digital, Electromagnetics, Instrumentation, Networks, Power, Signal Processing, Solid State Devices

Chemical • Reactions, Thermodynamics, Numerical Methods, Heat Transfer, Mass Transfer, Energy, Pollution, Controls, Design, Economics, Equipment, Safety, and Transport Phenomena

Industrial • Design, Econ, Statistics, Facilities, Cost Analysis, Computer Modeling, Ergonomics, Management, Systems, Manufacturing, Material Handling, Optimization, Production, Productivity, Queuing, Simulation, Quality Control, Quality Management, Performance

How to Use the NCEES Handbook

A copy of the NCEES Handbook may be mailed to you sometime after you register. It is the only reference allowed in the exam room, so you'll want to get to know it. If you're not sent a copy by your state board, you can order one from GLP (by way of our website or call 1-800-837-0201). Your first impression of this handbook may leave you overwhelmed! It has a tremendous amount of equations, figures, and information that may, at first glance, seem unfamiliar to you. Indeed, afternoon Discipline test info is mixed in with morning General info. In our review we have used the NCEES Handbook to compile summaries of the more important equations from the most significant General subject areas, fitting in with the Morning Session and our recommended Afternoon strategy—this is Appendix A at the back of this book. We have attempted to adapt our nomenclature throughout to that of the NCEES Handbook.

We strongly urge you to have the official NCEES Reference Handbook by your side the entire time you study, and use a highlighter to identify the equations you most often use. Initially, you can use our Equation Summary Sheets and your own best judgment to highlight the key equations in the general part of the Handbook.

Familiarize yourself with the location of these equations so you can quickly locate them during the actual exam in the new handbook that will be given to you at the test site. If you do not prepare this way, you might become lost in the handbook during the exam.

Strategies for Study

The following strategies will help you prepare for the FE General Sessions:

1. Focus your review on the eight subjects that have the most exam questions: Chemistry, Dynamics, Electrical, Fluids, Math, Solids, Statics, and Thermo.
2. *Quickly* review the remaining four major subject areas: Computers, Economics, Ethics and Materials Science.
3. Spend the majority of your time reviewing material with which you are *most* familiar.
4. Study throughout your review with the NCEES Handbook in a targeted, realistic manner.

You may receive a copy of the NCEES Handbook when you register for the exam (some states provide them), or you can order a copy from GLP by calling 1-800-837-0201. You will not be allowed to bring this handbook (or any other printed material) with you into the exam. However, a clean copy of this same handbook will be given to you upon entering the exam site. The new handbook is particularly confusing. It includes about 80 pages that cover the General subjects (a.m. and p.m.) and 45 pages that cover the Discipline-Specific subjects. The trouble is that the a.m. and p.m. subjects are intermixed! Thus, you'll find Heat Transfer (for the ME test) covered early in the Handbook, hidden in with most of the General topics. If you elect to take the General Test, boldly mark the pages that cover the general subjects. In our book, we have provided most of the same equations that you will find in the NCEES Handbook for the major a.m. and General Test p.m. subjects. Then search through the Handbook and highlight those important equations. This will allow you to familiarize yourself with the location of the equations you need most, and you will be able to quickly locate those equations in the Handbook during the exam. You will find this particularly helpful since the Handbook is filled with plenty of extraneous material. If you do not train yourself in this way, you may well find it difficult to quickly locate the appropriate equations during the exam. Being familiar with the Handbook is perhaps the most important detail you should attend to.

Outline of a Good Study Program for Quick Progress

For a senior in an engineering college who has a busy schedule, we suggest 8 weeks as an ideal study period. During those 8 weeks, you must be willing to perform a fairly concentrated study. You should set aside blocks of 3 hours at least 2 days a week. In any case, we recommend that you study no less than 4 weeks for the FE exam. For those of you who have been away from this material for a while, base the length of your study period on the number of years you have been away from school and your own memory capability.

Perform an initial review during the first half of your selected study period (for engineering seniors, 4 weeks). But if you are trying to get through your review quickly, only the Practice Problems that have been starred (*) should be worked and studied in our *Fundamentals of Engineering Review* (General Test).

Two days prior to exam day, review all subject areas briefly, using your highlighted handbook of equations and tables. Be sure you can quickly find the equations you will use most often in the *unmarked* booklet you will be given on exam day.

The day before the exam, relax and go to bed early. Do *not* cram or perform any panic studying. By now, you are as prepared as you can be for the exam.

The morning of your test, get up early and have a light, healthy breakfast (and maybe some coffee!). Arrive at the exam site at least 20 minutes early. You need to allow time for parking and getting settled—be sure to bring some change to pay for parking! During the one-hour lunch break, it is best to plan to meet with a friend who can help you relax and get refreshed for the Afternoon Session. If there are no restaurants nearby, bring a bag lunch and eat outside on the lawn somewhere. Get some fresh air. Do *not* spend the entire hour reviewing. Try not to talk to other test-takers about how it is going for them. This can easily induce either insecurity or false confidence. If you understand engineering principles and have prepared well, after the dust and sweat of the test day clears you'll find you've passed!

Pacing Yourself During the FE Exam

The problems in the morning session are, for the most part, unrelated. Consequently, *two minutes* (on the average) can be spent on each problem. This makes fast recall *essential*, as time does not allow you to contemplate various methods of solution. But each problem is only worth one-half point—so do not fall into the trap of spending too much time on any one problem from the morning session.

In the afternoon session, you can spend an average of 4 minutes per individual question. Many questions in the afternoon session are related to a common problem, divided into 2 to 6 sub-problems. Thus you may maintain a good pace and still spend up to 20-25 minutes solving a problem set that is more difficult.

Process of Elimination

Sometimes the best way to find the right answer is to look for the wrong ones and cross them out. On questions which are difficult for you, wrong answers are often much easier to find than right ones!

Answers are seldom given with more than three significant figures, and may be given with two significant figures. The choice *closest* to your own solution should be selected.

There is no penalty for *guessing*. Use the *process of elimination* when guessing at an answer. If only one answer is negative and three answers are positive, eliminate the one odd answer from your guess. Also, when in doubt, work backwards and eliminate those answers that you believe are untrue, and then guess. By using a combination of methods, you greatly improve your odds of answering correctly.

Should I Guess?

Leave the last ten minutes of each session for making educated guesses. **Do not leave any answers blank** on your answer key. A guess *cannot* hurt you, it can only help you. Your score is based on the number of questions you answer correctly. An incorrect answer does not harm your score.

Place a question mark beside choices you are uncertain of, but seem correct. If time prevents you from reworking that problem, you will have at least identified your best guess.

Difficult Problems

If at first glance you know that a certain problem will require much time and is exceptionally difficult for you, make your best guess, then *skip right over it*. Be sure to mark the problem in a unique manner (we suggest that you circle the problem number) so that if time permits, you may come back to it. (Note: it is not possible to return to the Morning Session problems in the Afternoon Session.)

When you are working through a problem and decide to move on due to difficulty, be sure to write down in your test booklet your notes and conclusions up to that point in case you have time to return to it. Then make your best guess and circle the problem number to return to if you have time.

If you feel you know how to work a difficult problem and could answer it with more time, identify it by circling the entire problem, not just the problem number—this identifies it as a "most likely" candidate for your set-aside ten minutes of "guess" time.

Time-Saving Tips

Once you determine your answer, *always write the letter corresponding to the correct answer in the margin of the test booklet* beside the question. At the end of the page, you can then transfer all the answers from that page to the answer key at once. This will save you considerable time and help you maintain concentration as well!

Cross out choices that you have eliminated in your test booklet on the problems you will return to. Otherwise, you will have to reread them as you make your last-ditch deliberations.

Write Out Your Work

Feel free to write all over the writing space provided for you in your test booklet. Do not hesitate to work out a problem, no matter how simple it may be. Doing as much work as you can on paper will ease your mind and leave it less "cluttered"—and it will help you if you need to return to a problem. You may think you're saving time, but your exam performance is *not* improved when you work problems in your head.

Bring a Calculator —Ideally, an HP 48GX

You must take a silent calculator (it may be preprogrammed) into the exam. A calculator is essential when solving many problems. In fact, with the exception of a couple of states, the premier engineering calculator, the HP 48GX, is allowed into the exam (check with your state board). This calculator is a hand computer which has hundreds of basic equations and constants preprogrammed. If you need extra help with exam calculations, the HP 48GX will be very useful. We at GLP offer this calculator for sale at substantial discounts. The GX, with its large memory and removable card-slot, is of particular importance because cards are made for it especially for the FE and PE exams —we have these cards available. We also offer a manual to guide you through the steps to using this calculator effectively for the FE Exam. (The manufacturer's manual is difficult to use for even basic operations.) Our engineering-oriented manual, called *Jump Start the HP 48G/GX*, will have you performing typical calculations in a very short time. Call us at 1-800-837-0201 for ordering or information, or browse our website with its easy-to-use order form at www.glpbooks.com.

English vs. SI Units

Some questions can be worked using either English units or SI units in the morning session only. We recommend that all test takers prepare using only SI units, although some of our review material uses English units since that is what is used in the course. The answers from the array of choices will be the same using either set of units. The problems in this book are mostly in SI units. The afternoon Civil Engineering Test will use both sets of units in some of the subjects areas (e.g., Soils). A table of conversion factors is presented in Appendix B of this book.

Recommended Materials for FE/EIT Preparation

1. *FE/EIT Discipline Reviews for Civil, Mechanical or Electrical*, (this book); GLP
2. *Fundamentals of Engineering Review—General*; a complete review; GLP
3. *FE/EIT Quick Prep*—for General sessions; a quick review; GLP
3. HP 48GX Programmable Calculator
4. *Jump Start the HP 48G/GX*, a how-to manual from GLP
5. NCEES FE/EIT Sample Exams, for each Discipline
6. NCEES Reference Handbook

All these resources, and more, are available from GLP by calling 1-800-837-0201.

Again... "Pass or Your Money Back"!

Did we forget to mention that with GLP you're guaranteed to pass? It's true! If you use our preps and fail the FE or PE exam, send us your name & address, receipt, book & CD, and copy of failure notice (within 30 days of notice)—and we will issue you a full refund.

State Boards of Registration Information

All State Boards of Registration administer the National Council of Engineering Examiners and Surveyors (NCEES) uniform examination. The dates of the exams cover a span of three days in mid-April and three days in late October. The specific dates are selected by each State Board. To be accepted to take the FE exam, an applicant must apply well in advance. For information regarding the specific requirements in your state, contact your State Board's office. If contact information has changed from what we have listed here, your correct State Board information can be obtained from the Executive Director of NCEES, PO Box 1686, Clemson, SC 29633-1686, ph 803-654-6824, or www.NCEES.org. Any comments relating to the exam or the Reference Handbook should be addressed to NCEES.

The answers to the following questions are answered alongside each listing:

1. *Do you provide an NCEES Handbook for all registrants?*
2. *Do you allow CE's, ME's, EE's, IE's and ChemE's to take the afternoon General Exam?*
3. *Are advanced calculators, such as the HP48GX, allowed?*

Columns: NCEES Handbook at signup? | Can all take Gen'l Exam? | Advanced calculators allowed?

Y Y Y ALABAMA: State Board of Licensure for Professional Engineers, P. O. Box 304451, Montgomery 36130-4451. Executive Secretary, Telephone: (334) 242-5568, engineer@dsmd.dsmd.state.al.us.

Y Y Y ALASKA: State Board of Registration for Engineers, Pouch D, Juneau 99811. Licensing Examiner, Telephone: (907) 465-2540, marcia_pappas@dced.state.ak.us, www.commerce.state.ak.us/occ/pael/htm.

N N Y ARIZONA: State Board of Technical Registration, 1990 W. Camelback Rd., Suite 406, Phoenix 85015. Executive Director, Telephone: (602) 255-4053, btrlvd@yahoo.com, www.btr.state.az.us.

Y Y 'Y' ARKANSAS: State Board of Registration for Professional Engineers and Land Surveyors, P. O. Box 3750, Little Rock 72203. Secretary-Treasurer, Telephone: (501) 682-2824, joe.clements@mail.state.ar.us. www.state.ar.us/pels. (No preprogrammed calculator cards allowed.)

Y Y Y CALIFORNIA: Board for Professional Engineers and Land Surveyors, 2535 Capitol Oaks Dr #300, Sacramento 95833. Executive Secretary, Telephone: (916) 263-2222, www.dca.ca.gov/pels.

'Y' Y Y COLORADO: State Board of Registration for Professional Engineers, 1560 Broadway, Suite 1370, Denver 80202. Program Administrator, Telephone: (303) 894-7788, www.dora.state.co.us/engineers. (Handbooks avail. while they last.)

N Y Y CONNECTICUT: State Board of Registration for Professional Engineers, The State Office Building, Rm 110, 165 Capitol Ave, Hartford 06106. Administrator, Telephone: (860) 713-6145.

State Boards of Registration

	NCEES Book?	Gen'l Exam?	Calculators?
DELAWARE: Association of Professional Engineers, 56 W. Main St, Suite 208, Christina 19702. Executive Secretary, Telephone: (302) 368-6708, peggy@dape.org, www.dape.org.	Y	Y	Y
DISTRICT OF COLUMBIA: Board of Registration for Professional Engineers, 941 N. Capitol St., OPLA Rm 2200 Washington 20002. Executive Secretary, Telephone: (202) 442-4320.	Y	Y	Y
FLORIDA: Board of Professional Engineers, 1208 Hays St., Tallahassee 32301-0755. Executive Director, Telephone: (850) 521-0500, board@fbpe.org, www.fbpe.org.	N	Y	Y
GEORGIA: State Board of Registration for Professional Engineers, 237 Coliseum Dr., Macon, 31217-3858. Executive Director, Telephone: (912) 207-1450, pels@sos.state.ga.us, www.sos.state.ga.us/ebd–pels.	Y	Y	Y
GUAM: Territorial Board of Registration for Professional Engineers, Architects and Land Surveyors, Department of Public Works, Government of Guam, P. O. Box 2950, Agana 96911. Chairman, Telephone: (671) 646-3115/3138.	?	?	?
HAWAII: State Board of Registration for Professional Engineers, P. O. Box 3469, Honolulu 96801. Executive Secretary, Telephone: (808) 586-2702.	N	N	Y
IDAHO: Board of Professional Engineers, 600 S. Orchard, Suite A, Boise 83705-1242. Executive Secretary, Telephone: (208) 334-3860, dcurtis@ipels.state.id.us, www.state.id.us/ipels.	Y	Y	Y
ILLINOIS: State Board of Professional Engineers, 320 West Washington, 3rd Fl, Springfield 62786. Unit Manager, Telephone: (217) 785-0820, question@dpr084.1.state.il.us, www.state.il.us.	N	Y	N
INDIANA: State Board of Registration for Professional Engineers, 302 W. Washington St., E034, Indianapolis 46204. Executive Director, Telephone: (317) 232-3902, www.ai.org/pla.	Y	Y	'Y'
IOWA: Engineering Examining Board, 1918 S.E. Hulsizer, Ankeny 50021. Executive Secretary, Tel: (515) 281-5602, jolene.schmitt@comm7,state.ia.us, www.state.ia.us/proflic.	N	Y	Y
KANSAS: State Board of Technical Professions, 900 Jackson, Suite 507, Topeka 66612. Executive Secretary, Telephone: (785) 296-3053, www.ink.org/public/ksbtp.	Y	N	Y
KENTUCKY: State Board of Licensure for Professional Engineers, 160 Democrat Dr., Frankfort 40601. Executive Director, Telephone: (502) 573-2680, larry.perkins@mail.state.ky.us, www.kyboels.	Y	Y	Y

12 Passing the FE/EIT Exam

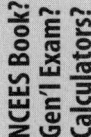

NCEES Book?	Gen'l Exam?	Calculators?	
Y	Y	Y	**LOUISIANA:** State Board of Registration for Professional Engineers, 10500 Coursey Blvd Suite 107, Baton Rouge. Executive Secretary, Telephone: (225) 295-8522, www.lapels.com.
Y	Y	Y	**MAINE:** State Board of Registration for Professional Engineers, 92 State House, Station, Augusta 04333-0092. Secretary, Telephone: (207) 287-3236, pengineers@ctel.net, www.professionalsmaineusa,com.
N	Y	Y	**MARYLAND:** Board for Professional Engineers, 500 N. Calvert St, Rm 308, Baltimore 21202-3651. Executive Secretary, Telephone: (410) 230-6322, dmatricciani@dllr.state.md.us, www.dllr.state.md.us.
N	Y	Y	**MASSACHUSETTS:** State Board of Registration of Professional Engineers, 239 Canseway St, Boston 02114. Secretary, Telephone: (617) 727-3074, marie.e.deveau@state.ma.us, www.state.ma.us/reg.
'Y'	Y	Y	**MICHIGAN:** Board of Professional Engineers, P. O. Box 30018, Lansing 48909. Administrative Secretary, Telephone: (517) 241-9253, jack.sharpe@cis.state.mi.us. (NCEES Handbook provided upon signup if requested.)
Y	N	N	**MINNESOTA:** State Board of Registration for Engineers, 85 E. 7th Pl, Suite 160, St. Paul 55101. Executive Secretary, Telephone: (651) 296-2388, sheri.lindemann@state.mn.us, www.aelslagid.state.mn.us.
Y	Y	Y	**MISSISSIPPI:** State Board of Registration for Professional Engineers, P. O. Box 3, Jackson 39205. Executive Director, Telephone: (601) 359-6160, information@pepls.state.ms.us, www.peplsstate.ms.us.
N	Y	Y	**MISSOURI:** Board of Professional Engineers, P. O. Box 184, Jefferson City 65102. Executive Director, Telephone: (573) 751-0047, moapels@mail.state.mo.us, www.ecodev.state.mo.us/pr/apels.
N	Y	Y	**MONTANA:** State Board of Professional Engineers and Land Surveyors, Department of Commerce, 111 N. Jackson, P. O. Box 200513, Helena 59620-0513. Administrative Secretary, Telephone: (406) 444-1667, compolpel@state.my.us, www.com.state.mt.us/license/POL/pol_boards.
Y	N	Y	**NEBRASKA:** State Board of Professional Engineers, 301 Centennial Mall South, 6th Fl, Lincoln 68508. Executive Director, Telephone: (402) 471-2021, execdir@nol.org, www.nol.org/home/NBOP.
Y	Y	Y	**NEVADA:** State Board of Professional Engineers, 1755 East Plum Lane, Ste. 135, Reno 89502. Executive Secretary, Telephone: (775) 688-1231, nevengsur@natinfo.net, www.state.nv.us/BOE.

State Boards of Registration 13

	NCEES Book?	Gen'l Exam?	Calculators?

NEW HAMPSHIRE: State Board of Professional Engineers, 57 Regional Drive, Concord 03301. Executive Secretary, Telephone: (603) 271-2219, llavertu@nhsa.state.nh.us, www.state.nh.us/jtboard/home. Y Y Y

NEW JERSEY: State Board of Professional Engineers and Land Surveyors, P. O. Box 45015, Newark 07101. Executive Secretary-Director, Telephone: (973) 504-6460. Y Y Y

NEW MEXICO: State Board for Professional Engineers, 1010 Marquez Pl., Santa Fe 87501. Secretary, Telephone: (505) 827-7561, amanda.lopez@state.nm.us, www.state.nm.us/pepsboard. Y Y N

NEW YORK: State Board for Engineering, Cultural Education Center, Rm 3019, Albany 12230. Executive Secretary, Telephone: (518) 474-3846, enginbd@mail.nysed.gov, www.nysed.gov/prof/pe. N Y Y

NORTH CAROLINA: State Board of Professional Engineers, 310 W Millbrook Rd, Raleigh 27609. Executive Secretary, Telephone: (919) 841-4000, ncboard@ncbels.org, www.ncbels.org. Y Y Y

NORTH DAKOTA: State Board of Registration for Professional Engineers, P. O. Box 1357, Bismarck 58502. Executive Secretary, Telephone: (701) 258-0786. Y Y 'N'

OHIO: State Board of Registration for Professional Engineers, 77 S. High St., 16th Fl., Columbus 43266-0314. Executive Secretary, Telephone: (614) 466-3650, mjacob@mail.peps.state.oh.us, www.peps.state.oh.us. Y Y Y

OKLAHOMA: State Board of Registration for Professional Engineers, 201 N.E. 27th Street, Rm 120, Oklahoma City, 73105-2788. Executive Secretary, Telephone: (405) 521-2874, www.okpels.org. Y Y Y

OREGON: State Board of Engineering Examiners, Department of Commerce, 728 Hawthorne Ave NE, Salem 97301. Executive Secretary, Telephone: (503) 362-2666, grahame@osbeels.org, www.osbeels.org. Y Y Y

PENNSYLVANIA: State Registration Board for Professional Engineers, P. O. Box 2649, Harrisburg 17105-2649. Administrative Assistant, Telephone: (717) 783-7049, engineer@pados.dos.state.pa.us, www.dos.state.pa.us/bpoa/engbd. N Y Y

PUERTO RICO: Board of Examiners of Engineers, P. O. Box 9023271, San Juan 00907-3271. Director, Examining Boards, Telephone: (728) 722-4816. ? ? ?

RHODE ISLAND: Board of Registration for Professional Engineers, 1 Capitol Hill, 3rd Fl, Providence 02908, Administrative Assistant, Ph: (401) 222-2565. N N Y

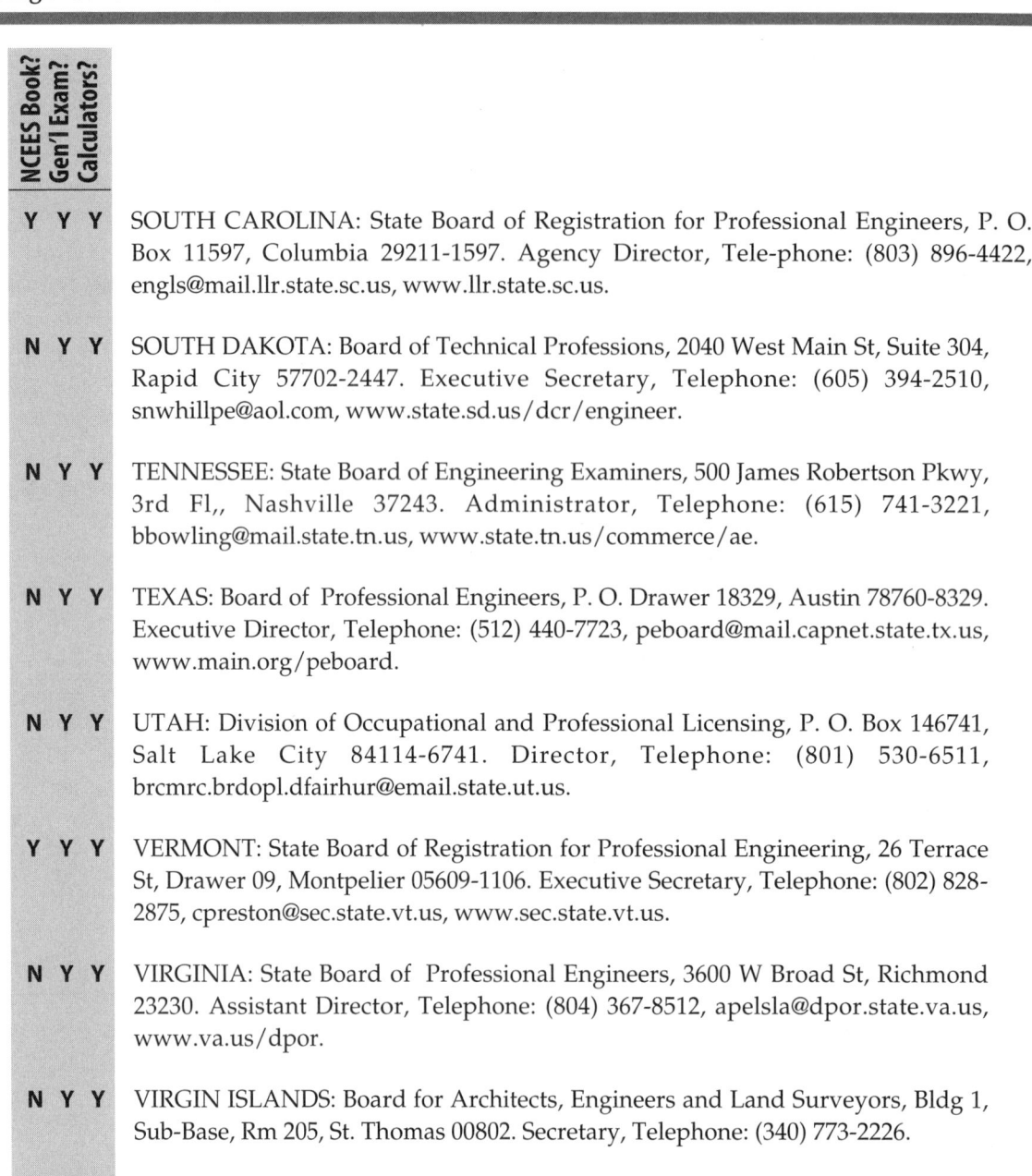

14   Passing the FE/EIT Exam

NCEES Book?	Gen'l Exam?	Calculators?	
Y	Y	Y	SOUTH CAROLINA: State Board of Registration for Professional Engineers, P. O. Box 11597, Columbia 29211-1597. Agency Director, Tele-phone: (803) 896-4422, engls@mail.llr.state.sc.us, www.llr.state.sc.us.
N	Y	Y	SOUTH DAKOTA: Board of Technical Professions, 2040 West Main St, Suite 304, Rapid City 57702-2447. Executive Secretary, Telephone: (605) 394-2510, snwhillpe@aol.com, www.state.sd.us/dcr/engineer.
N	Y	Y	TENNESSEE: State Board of Engineering Examiners, 500 James Robertson Pkwy, 3rd Fl,, Nashville 37243. Administrator, Telephone: (615) 741-3221, bbowling@mail.state.tn.us, www.state.tn.us/commerce/ae.
N	Y	Y	TEXAS: Board of Professional Engineers, P. O. Drawer 18329, Austin 78760-8329. Executive Director, Telephone: (512) 440-7723, peboard@mail.capnet.state.tx.us, www.main.org/peboard.
N	Y	Y	UTAH: Division of Occupational and Professional Licensing, P. O. Box 146741, Salt Lake City 84114-6741. Director, Telephone: (801) 530-6511, brcmrc.brdopl.dfairhur@email.state.ut.us.
Y	Y	Y	VERMONT: State Board of Registration for Professional Engineering, 26 Terrace St, Drawer 09, Montpelier 05609-1106. Executive Secretary, Telephone: (802) 828-2875, cpreston@sec.state.vt.us, www.sec.state.vt.us.
N	Y	Y	VIRGINIA: State Board of Professional Engineers, 3600 W Broad St, Richmond 23230. Assistant Director, Telephone: (804) 367-8512, apelsla@dpor.state.va.us, www.va.us/dpor.
N	Y	Y	VIRGIN ISLANDS: Board for Architects, Engineers and Land Surveyors, Bldg 1, Sub-Base, Rm 205, St. Thomas 00802. Secretary, Telephone: (340) 773-2226.
Y	Y	Y	WASHINGTON: State Board of Registration for Professional Engineers and Land Surveyors, P. O. Box 9649, Olympia 985047-9649 Executive Secretary, Telephone: (360) 753-6966, engineers@dol.wa.gov, www.wa.gov/dol/bpd/engfront.
N	Y	Y	WEST VIRGINIA: State Board of Registration for Professional Engineers, 608 Union Building, Charleston 25301-2703. Executive Director, Telephone: (304) 558-3554.
Y	Y	Y	WISCONSIN: State Examining Board of Professional Engineers, P. O. Box 8935, Madison 53708-8935. Administrator, Telephone: (608) 266-5511, dorl@mail.state.wi.us, www.badger.state.wi.us/agencies/drl.
Y	Y	Y	WYOMING: State Board of Examining Engineers, 2424 pioneer Ave, Suite 400, Cheyenne 82001. Secretary-Accountant, Telephone: (307) 777-6155, cturk@wyoming.com, www.wrds.uwyo.edu/wrds/borpe/borpe.

The FE/EIT Review for Electrical Engineering

edited by Merle C. Potter

Chapters

1. Electromagnetic Fields .... Zapp
2. Power .... Wiitanen
3. Communications and Signal Processing .... Zapp
4. Solid State .... Ho
5. Computer Engineering .... Tummala

Most of the problems on the Electrical Engineering test of the afternoon session can be worked using pages 99 to 105 of the NCEES Handbook, 3rd ed. Many of the problems, however, on the afternoon Electrical Engineering test will come from the material in pages 1 to 83 of the NCEES Handbook, material used for the General tests of the morning and afternoon sessions. Since all engineers must take the General morning exam, it is assumed that you are familiar with that material. You must realize, however, that you will have to make reference to equations, tables, and charts located in those first 83 pages of the NCEES Handbook when you are working problems in the afternoon Electrical Engineering session. The general engineering sections that you should be particularly familiar with

Overview of the FE/EE Topics

What You Need to Know About the NCEES Handbook

16   Overview of FE/Electrical Review

are the sections on Mathematics, Electric Circuits, and Control Systems; a review of problems on Control Systems can be found in the Mechanical Engineering part of this book. A review of General engineering material is found in our *Fundamentals of Engineering*, a review book written specifically for the FE/EIT exam.

Do I Have to Take the Discipline Test?

If, after you study this Electrical Engineering review section, and take the Electrical Engineering Discipline Practice Exam, you feel that you could more easily pass the General exam of the afternoon session, you should register to take the afternoon General test. If you have already registered for the Electrical Engineering afternoon test, you may contact your state board and request a change. Make sure your state board recognizes a *pass* on the General afternoon test for your particular discipline before you make that decision; the vast majority of the state boards will recognize a pass on any of the afternoon tests by any registrant.

Why this Review is Focused on the NCEES Handbook

In order to provide a very efficient, yet effective review, this DS Review has been designed to review only the material associated with the formulas and figures presented in the NCEES Handbook. The Handbook is the only material allowed in the exam. Thus it makes an efficient base to study around. It is assumed that you have a copy of the NCEES Handbook, 3rd ed. If you do not, you can obtain one from Great Lakes Press by calling 1-800-837-0201.

GLP Offers a Concise A.M. Session Review & Other Study Aids

We also have an effective review for the General tests, both morning and afternoon, entitled *Fundamentals of Engineering*. It is used at hundreds of engineering schools across the nation. We also distribute the HP 48GX calculator, which is pre-programmed with many of the equations you'll need when working problems on the FE/EIT exam. Also, we've developed *Jump Start Your HP 48G/GX*, a 240-page book designed to efficiently help engineers get the most out of the HP48G/GX. You can order these items by calling 1-800-837-0201.

1. Electromagnetic Fields

by Roland Zapp

The equations used to solve problems involving electromagnetic fields are listed on page 99 of the NCEES Handbook, 3rd ed. Related information may be needed from pages 68 to 71 of the Handbook.

Example 1.1

The electric field for a dipole located at the origin of a spherical coordinate system is given by

$$\mathbf{E} = \frac{2k\cos\theta\, \mathbf{a}_r}{4\pi\varepsilon_0 r^3} + \frac{k\sin\theta\, \mathbf{a}_\theta}{4\pi\varepsilon_0 r^3}$$

1. A thin spherical surface of radius $r = 25$ cm is located at the origin of the given spherical coordinate system. It was determined that Maxwell's equations yield only the dipole field given. The potential on the surface of the sphere is:
   (A) zero
   (B) constant
   (C) $A/(r^2\cos\theta)$
   (D) $V_0 \cos\theta$ where V_0 is a constant

2. A Gaussian surface of radius r_0 encloses the dipole field given. The surface integral $\oint_A \mathbf{E} \cdot d\mathbf{A}$ gives a value of:
   (A) $r_0 k$
   (B) $r_0 k / 2$
   (C) k (where $k = Qd$ in the limit $d \to 0$)
   (D) None of the above

3. Applications of Maxwell's equations
$$\nabla \times \mathbf{E} = -\frac{\partial \mathbf{B}}{\partial t} \quad \text{and} \quad \nabla \cdot \mathbf{E} = \frac{\rho}{\varepsilon_o}$$
   to the given dipole field would yield
   (A) $\mathbf{B} = 0, \rho = 0$ everywhere
   (B) The curl and divergence are indeterminate
   (C) The equations are well behaved except at the origin
   (D) None of the above

Ch. 1 / Electromagnetic Fields

Solutions:

1. **D** $V_0 \cos\theta$.

    General potential of a dipole is $\dfrac{P_0 \cos\theta}{4\pi\varepsilon_0 r^2}$ which at r = constant gives $K\cos\theta$.

2. **D** By Gauss's Law, the surface integral of **E** *over* any surface gives the charge enclosed. Since a dipole represents a zero net charge, the answer is zero.

3. **C** The curl and divergence are zero everywhere except at $r = 0$, where the derivatives are undefined.

Example 1.2

Given the electric field:
$$\mathbf{E} = 2x^2 \mathbf{a}_x + 2y^2 \mathbf{a}_y + 2z^2 \mathbf{a}_z$$

1. Select one of the following:
    - (A) $\nabla \cdot \mathbf{E} \neq 0$
    - (B) $\oint_L \mathbf{E} \cdot d\mathbf{L} = 0$
    - (C) $\oint_S \mathbf{E} \cdot d\mathbf{s} \neq 0$
    - (D) All of the above

2. The work done in moving a negative charge $-|Q|$ from (0,0,0) to (1,1,1) in the electric field described is:
    - (A) zero
    - (B) two $|Q|$
    - (C) four $|Q|$
    - (D) None of the above

3. The potential associated with the given electric field is:
    - (A) $V = -4(x+y+z)$
    - (B) $V = -2(x^3 + y^3 + z^3)/3$
    - (C) $V = -2(x^3 y^3 z^3)/3$
    - (D) None of the above

Solutions:

1. **D** All of the above.

2. **B** Two $|Q|$.

3. **B** $V = -2(x^3 + y^3 + z^3)/3$.
    Note $\mathbf{E} = -\nabla V$. Can also get the answer to Prob. 2 by noting $W = QV$.

Example 1.3

Assume an electric flux density of $\mathbf{D} = 2x^2\mathbf{a}_x$. The total charge enclosed in a cube of side = 2 located in the first quadrant with $x = y = z = 0$ as one of its corners is:

(A) zero
(B) four
(C) sixteen
(D) thirty-two

Solution:

$$\nabla \cdot \mathbf{D} = \rho. \quad \int \rho \, dv = Q = 32.$$

The solution is **D**.

Example 1.4

A circular wire ring of radius a centered at the origin of a cylindrical coordinate system has a total positive charge Q uniformly distributed over the wire. The electric field at a location $z = z_0$ on the z-axis is given by:

(A) $\dfrac{Q}{4\pi\varepsilon_0(a^2 + z_0^2)}\mathbf{a}_z$

(B) $\dfrac{Q}{4\pi\varepsilon_0(a^2 + z_0^2)}\mathbf{a}_r$

(C) $\dfrac{Q}{4\pi\varepsilon r^2}\mathbf{a}_r$

(D) None of the above.

Solution:

By simply applying integration to the point charge field $\mathbf{E} = \dfrac{Q}{4\pi\varepsilon r^2}\mathbf{a}_r$ the field becomes

$$\mathbf{E} = \frac{Qz_0}{4\pi\varepsilon_0(a^2 + z_0^2)^{3/2}}\mathbf{a}_z$$

The solution is **D**.

Example 1.5

A discontinuity on a transmission line is detected by launching a known step voltage and observing the echo. This is done as shown below.

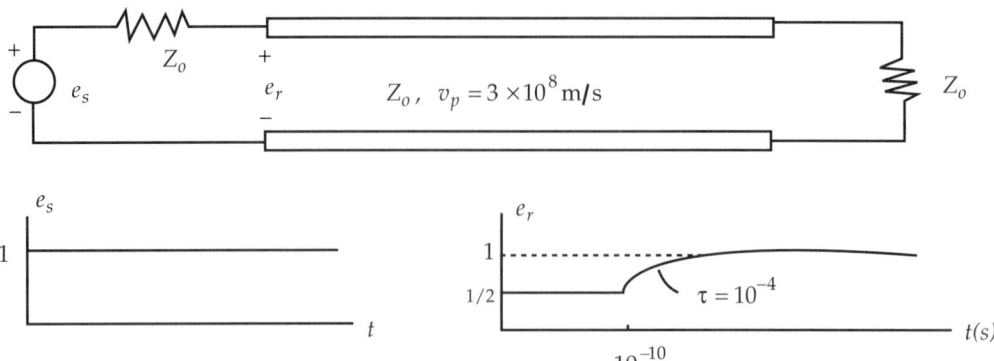

The following can be determined from this exercise:

(A) A discontinuity exists 3 cm from the line input.
(B) An open circuit exists on the line.
(C) An inductor of value 5×10^{-3} is present on the line.
(D) None of the above.

Solution:

A capacitive discontinuity exists 1.5 cm down the line (with $C = 2\mu F$).

The answer is **D**.

Example 1.6

Current flows axially with uniform density $\mathbf{J} = J_0 \mathbf{a}_z$ amp/m^2 along a cylindrical conductor of radius a and length L with uniform conductivity σ_0, by the application of a potential difference V_0 between the ends of the conductor. Determine the magnetic field $\mathbf{H}$ inside the cylinder.

(A) $\mathbf{H} = \dfrac{J_0 r}{2} \mathbf{a}_\phi$

(B) $\mathbf{H} = \dfrac{J_0 V_0}{2Lr} \mathbf{a}_\phi$

(C) $\mathbf{H} = \dfrac{J_0}{2\pi r} \mathbf{a}_\phi$

(D) $\mathbf{H} = \dfrac{J_0 a^2}{2r} \mathbf{a}_\phi$

Solution:

Using Maxwell's equation, $\oint \mathbf{H} \cdot d\mathbf{L} = \int \mathbf{J} \cdot d\mathbf{s}$ and symmetry.

The answer is **A**.

Example 1.7

The power dissipated by the cylinder of Example 1.6 is:

(A) $P_d = J_0^2 / \sigma_0$

(B) $P_d = \dfrac{V_0^2 \sigma_0 a^2 \pi}{L}$

(C) $P_d = \dfrac{L J_0^2 a^2 \pi}{2\sigma_0}$

(D) None of the above

Solution:

$$P_d = \int \mathbf{J} \cdot \mathbf{E} dv = \frac{L}{\sigma_0} \int_{\varphi=0}^{2\pi} \int_{r=0}^{a} J_0^2 r \, dr \, d\varphi = \frac{L}{\sigma_0} J_0^2 a^2 \pi$$

where $J_0 = \sigma_0 \dfrac{V_0}{L}$.

The answer is **B**.

Example 1.8

The electric field **E** in free space is given by:

$$\mathbf{E} = \mathbf{a}_y E_0 \sin(\omega t - kz)$$

Determine the magnetic field **B** associated with this electric field.

(A) $\mathbf{B} = \mathbf{a}_x [-\dfrac{E_0 k}{\omega} \cos(\omega t - kz)]$

(B) $\mathbf{B} = \mathbf{a}_x [-\dfrac{E_0 k}{\omega} \sin(\omega t - kz)]$

(C) $\mathbf{B} = \mathbf{a}_y [-\dfrac{E_0 k}{\omega} \cos(\omega t - kz)]$

(D) $\mathbf{B} = \mathbf{a}_y [-\dfrac{E_0 k}{\omega} \sin(\omega t - kz)]$

Solution:

This is solved using $\nabla \times \mathbf{E} = -\dfrac{\partial \mathbf{B}}{\partial t}$. The answer is **B**.

2. Electric Power

by Dennis Wiitanen

Page 99 of the NCEES Handbook provides some information on electric power, listing the fundamental relationships for ac Machines and balanced three phase systems. In addition, a knowledge of ac circuits as outlined on pages 70 and 71 is required to work the basic problems in power.

2.1 Balanced Three Phase Systems

Three phase systems are employed because of ease of generation and transmission efficiency. Solving balanced three phase problems requires the application of the equations on page 99 in conjunction with the ac circuits relationships on pages 70 and 71.

Example 2.1 Wye Connected Load

A three phase load of wye connected impedances of 4.0 + *j*3.0 ohms each is supplied by a balanced three phase source with a line voltage of 207.8 volts.

1. What is the magnitude of the voltage across each impedance?
    (A) 120 V
    (B) 207.8 V
    (C) 360 V
    (D) 440 V

2. What is the power factor of the load?
    (A) 0.20
    (B) 0.40
    (C) 0.80
    (D) 36.8

3. What is the magnitude of the line current?
    (A) 10 A
    (B) 24 A
    (C) 38 A
    (D) 41.6 A

4. What is the real power delivered to the load?
   (A) 2.30 kW
   (B) 2.88 kW
   (C) 3.99 kW
   (D) 6.91 kW

5. What is the reactive power delivered to the load?
   (A) 1.73 kVAR
   (B) 2.88 kVAR
   (C) 5.18 kVAR
   (D) 8.64 kVAR

Solutions:

1. **A** $|V_P| = |V_L|/\sqrt{3} = 207.8/\sqrt{3} = 120.0$

2. **C** p.f. $= \cos \theta_p = \cos(\tan^{-1} 3/4) = .80$

3. **B** $|I_P| = |V_P|/|Z_P|$; for a wye system, $|V_P| = |V_L|/\sqrt{3}$ and $I_P = I_L$.
   Therefore, $|I_L| = (|V_L|/\sqrt{3})/|Z_P|$; also, $|Z_P| = (R^2 + X^2)^{0.5}$
   Finally,
   $$|I_L| = 207.8/\sqrt{3(4.0^2 + 3.0^2)} = 24.0 \text{ A}$$

4. **D** $P = \sqrt{3}|V_L||I_L|\cos \theta_p = \sqrt{3}(207.8)(24)\cos(\tan^{-1} 3/4) = 6910 \text{ W} = 6.91 \text{ kW}$

5. **C** $Q = \sqrt{3}|V_L||I_L|\sin \theta_p = \sqrt{3}(207.8)(24)\sin(\tan^{-1} 3/4) = 5{,}180 \text{ W} =$
   5.18 kVAR

Example 2.2 General Load Specification

A balanced three phase load supplied by a 460 volt balanced three phase source draws 50 kVA at an 0.9 lagging power factor.

1. What is the real power delivered to the load?
   (A) 12.3 kW
   (B) 15.0 kW
   (C) 32.8 kW
   (D) 45.0 kW

2. What is the reactive power delivered to the load?
   (A) 21.8 kVAR
   (B) 32.6 kVAR
   (C) 45.0 kVAR
   (D) 50.0 kVAR

3. What is the magnitude of the line current the load draws from the source?
   - (A) 12.6 A
   - (B) 19.1 A
   - (C) 41.6 A
   - (D) 62.8 A

4. If the load were represented by a set of delta connected impedances, what would be the required impedance per phase?
   - (A) $1.41 \angle 25.8° \, \Omega$
   - (B) $2.44 \angle 25.8° \, \Omega$
   - (C) $4.23 \angle 25.8° \, \Omega$
   - (D) $12.7 \angle 25.8° \, \Omega$

5. If the load were represented by a set of wye connected impedances, what would be the required impedance per phase?
   - (A) $1.41 \angle 25.8° \, \Omega$
   - (B) $2.44 \angle 25.8° \, \Omega$
   - (C) $4.23 \angle 25.8° \, \Omega$
   - (D) $7.33 \angle 25.8° \, \Omega$

Solutions:

1. **D** $VA = P + jQ$; $P = \text{Re}\{VA\} = VA \cos \theta_P$; p.f. $= \cos \theta_P$; therefore

   $P = VA(\text{p.f.}) = 50{,}000(.9) = 45{,}000 \text{ W} = 45.0 \text{ kW}$

2. **A** $VA = P + jQ$; $Q = \text{Im}\{VA\} = VA \sin \theta_P$; p.f. $= \cos \theta_P$; so

   $Q = VA(\sin \cos^{-1}\text{p.f.}) = 50{,}000(\sin (\cos^{-1}.9)) = 50{,}000(\sin \pm 25.8°)$ (positive angle for lagging p.f.)

   Therefore,

   $Q = 50{,}000(\sin 25.8°) = 21{,}800 \text{ VAR} = 21.8 \text{ kVAR}$

3. **D** $VA = \sqrt{3}\,|V_L|\,|I_L|$; which gives

   $|I_L| = VA/(\sqrt{3}\,|V_L|) = 50{,}000/(\sqrt{3} \times 460) = 62.8 \text{ A}$

4. **D** $|Z_P| = |V_P|/|I_P|$; in a delta load $V_P = V_L$ and $|I_P| = |I_L|/\sqrt{3}$

   Therefore

   $|Z_\Delta| = |V_L|/(|I_L|/\sqrt{3}) = 460/(62.8/\sqrt{3}) = 12.7$ ohm

   Also, p.f. $= \cos \theta_P$, so $\theta_P = \cos^{-1}(\text{p.f.}) = \cos^{-1}(.9) = \pm 25.8°$ (positive angle for lagging p.f.) so

   $Z_\Delta = |Z_\Delta| \angle 25.8° = 12.7 \angle 25.8°$

5. **C** $|Z_P| = |V_P|/|I_P|$; in a wye load $|V_P| = |V_L|/\sqrt{3}$ and $I_P = I_L$.

   Therefore $|Z_Y| = |V_L|/\sqrt{3}/|I_L| = (460/\sqrt{3})/62.8 = 4.23$ ohm.

Also, p.f. = cos θ_p, so θ_p = cos^{-1} (p.f.) = cos^{-1} (.9) = ±25.8° (positive angle for lagging p.f.) so

$Z_Y = |Z_Y| \angle 25.8° = 4.23 \angle 25.8°$

2.2 AC Machines

The two dominant ac machines in use are the synchronous machine and the induction machine. The synchronous machine is used primarily in the generation of electric power, while most applications converting electrical to mechanical power employ induction motors. The primary equations relating shaft speed to applied frequency are given on page 99.

Example 2.3 Induction Motor

A 6-pole induction motor operating under load drawing 9.35 kW from a 460 volt three phase source delivers 10.0 horsepower at a slip of 5.0 percent.

1. What is the synchronous speed of the motor?
   (A) 1200 rpm
   (B) 1800 rpm
   (C) 3258 rpm
   (D) 7540 rpm

2. What is the shaft speed of the motor?
   (A) 900 rpm
   (B) 1140 rpm
   (C) 1200 rpm
   (D) 1710 rpm

3. What torque is being delivered to the load?
   (A) 0.56 N·m
   (B) 1.04 N·m
   (C) 9.76 N·m
   (D) 62.5 N·m

4. What is the efficiency of the motor at this load?
   (A) zero
   (B) 50%
   (C) 80%
   (D) 91%

Solutions:

1. **A** $n_s = 120\, f/P = 120(60)/6 = 1{,}200$ rpm

2. **B** $n_r = (1-s)n_s = (1-s)120\, f/P = (1-0.05)120(60)/6 = 1140$ rpm

3. **D** $T = P(\text{watts})/\omega(\text{rad/sec})$; 1 h.p. = 746 watts

    Therefore,

    $$T = \frac{10 \times 746}{1140 \times 2\pi/60} = 62.5 \text{ N·m}$$

4. **C** Efficiency $= P_{out}/P_{in} = (746)(10.0)/9350 = 0.80 = 80\%$

3. Communications and Signal Processing

by H. Roland Zapp

The material in the NCEES Reference Handbook used to solve problems in Communications and Signal Processing is found on page 99, with some continuation on page 100. There may also be some reference required to information contained on pages 68 through 71 of the Handbook.

Example 3.1

A linear-time-invariant (LTI) filter with impulse response
$$x(t) = e^{-t} \quad t \geq 0$$
$$= 0 \quad t < 0$$
has an input
$$x(t) = e^{-2t} \quad t \geq 0$$
$$= 0 \quad t < 0$$

1. The output response is given by:

    (A) $y(t) = e^{-2t} \quad t \geq 0$

    (B) $y(t) = e^{-t} \quad t \geq 0$

    (C) $y(t) = 2e^{-t}(1 - e^{-t}) \quad t \geq 0$

    (D) $y(t) = e^{-t}(1 - e^{-t}) \quad t \geq 0$

2. The LTI filter has a system function given by:

    (A) $H(\omega) = \dfrac{1}{1 + j\omega} \quad \omega \geq 0$

    (B) $H(\omega) = \dfrac{1}{1 + \omega^2} - j\dfrac{\omega}{1 + \omega^2} \quad \omega \leq 0$

    (C) $|H(\omega)| = \dfrac{\sqrt{1 + \omega^2}}{1 + \omega^2} \quad -\infty \leq \omega \leq \infty$

    (D) None of the above

Solutions:

1. **D** Using $y(t) = \int_0^t x(\tau)h(t-\tau)d\tau$ yields $y(t) = e^{-t}(1-e^{-t})$.

2. **C** Based on the Fourier transform:
$$H(\omega) = \frac{1}{1+j\omega} = \frac{1-j\omega}{1+\omega^2} \quad -\infty \leq \omega \leq \infty$$

$$\therefore |H(\omega)| = \frac{\sqrt{1+\omega^2}}{1+\omega^2} \quad -\infty \leq \omega \leq \infty$$

Example 3.2

Find the impulse response $h(t)$ of a filter which yields the output to the input:

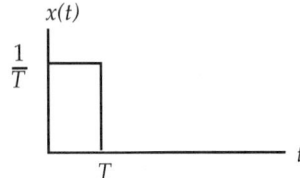
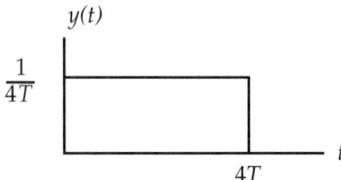

(A) $h(t) = 1 \quad 0 \leq t \leq 4T$

(B) $h(t) = 4 \quad 0 \leq t \leq 4T$

(C) $h(t) = \frac{1}{4T^2}t \quad 0 \leq t \leq T$

$\quad\quad = \frac{1}{4T^2} \quad T \leq t \leq 4T$

$\quad\quad = \frac{1}{4T^2}(5T-t) \quad 4T \leq t \leq 5T$

(D) None of the above

Solution:

By convolution, a train of 4 impulses, separated by T and of amplitude 1/4 will give the desired output.

The answer is **D**.

Example 3.3

The periodic impulse train shown is passed through the idealized low-pass filter shown. What is the average output power?

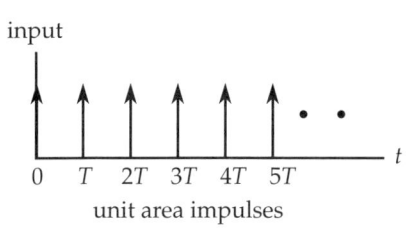

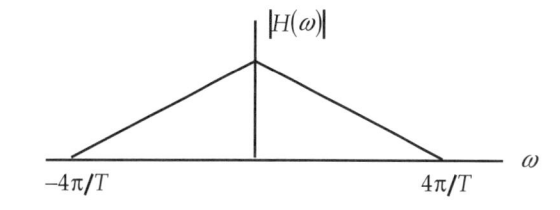

input

$0 \quad T \quad 2T \quad 3T \quad 4T \quad 5T$

unit area impulses

$|H(\omega)|$ with base from $-4\pi/T$ to $4\pi/T$

(A) $\langle P \rangle = \dfrac{5}{4T^2}$

(B) $\langle P \rangle = \dfrac{1}{2T^2}$

(C) $\langle P \rangle = \dfrac{3}{2T^2}$

(D) None of the above

Solution:

Input $= \sum_{k=-\infty}^{\infty} c_k e^{jk\omega_0 T}$ where $\omega_0 = \dfrac{2\pi}{T}$ and $c_k = \dfrac{1}{T}$ for all k.

Filter allows only c_0, c_{-1}, c_{+1} or $c_0^2 + 2\left[\dfrac{c_1}{2}\right]^2 = \dfrac{3}{2T^2}$.

The answer is **C**.

Example 3.4

The Fourier series $x(t) = \sum c_n e^{j2\pi f_0 n t}$ for the signal shown is:

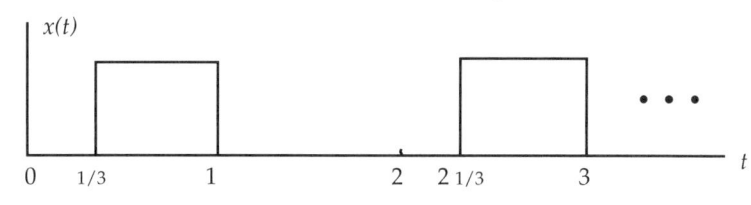

$x(t)$ with pulses from $1/3$ to 1, and $2\ 1/3$ to 3, ...

(A) $c_n = \dfrac{2e^{-j2\pi n/3}}{\pi n} \sin\left(\dfrac{\pi n}{3}\right)$

(B) $c_n = \dfrac{2}{3} \operatorname{sinc}\left(\dfrac{\pi n}{3}\right) e^{-2\pi n f_0/3}$

(C) $c_n = \operatorname{sinc}\left(\dfrac{\pi n}{3}\right) e^{-4\pi n f_0/3}$

(D) None of the above

Solution:

Use a Fourier series expansion. See page 17 of the NCEES Handbook.

The answer is **A**.

Example 3.5

The energy contained in an aperiodic signal is given by:

$$E = \int_{-\infty}^{\infty} f^2(t)dt$$

If $f(t) = \dfrac{\sin(t-\tau)}{t-\tau}$ the energy is

(A) infinite
(B) one
(C) π
(D) 2π

Solution:

Using the Parseval power theorem (or energy theorem):

$$\int_{-\infty}^{\infty} f^2(t)dt = \int_{-\infty}^{\infty} |F(f)|^2 df$$

$$= \int_{-1/2\pi}^{1/2\pi} \pi^2 df = \pi$$

The answer is **C**.

Example 3.6

A tone modulating AM signal is given by:

$$A_c[1 + \cos \omega_m t]\cos \omega_c t$$

where $\omega_c = 600\pi \times 10^3$ and $\omega_m = 6\pi \times 10^3$.

1. This will have positive frequency spectral components located at:
   (A) 300 KHz and 303 KHz
   (B) 300 KHz and 297 KHz
   (C) 297 KHz and 303 KHz
   (D) None of the above

2. The AM signal is transmitted by USSB-SC (upper single side band suppressed carrier). The expression for this signal is:
   (A) $\dfrac{A_c}{2}\cos(606 \times 10^3 \pi t)$
   (B) $A_c \cos(6 \times 10^3 \pi t)\cos(600 \times 10^3 \pi t)$
   (C) $A_c \cos(600 \times 10^3 \pi t)$
   (D) None of the above

Solutions:

1. **D** Components will be located at all 3 frequencies: 297, 300, 303 KHz.
2. **A** Only the upper spectral frequency is transmitted.

Example 3.7

Consider the tone-modulated FM signal

$$x(t) = \cos(2 \times 10^8 \pi t + \sin 400 \pi t)$$

applied to an ideal band pass filter of bandwidth 500 Hz and centered at 10^8 Hz. The output from this filter has sinusoidal components at frequencies:

(A) 10^8 Hz plus $(10^8 + 200)$ Hz plus $(10^8 + 400)$ Hz

(B) 10^8 Hz plus $(10^8 - 200)$ Hz plus $(10^8 - 400)$ Hz

(C) 10^8 Hz plus $(10^8 - 200)$ Hz plus $(10^8 + 200)$ Hz

(D) None of the above

Solution:

The FM spectrum contains the carrier plus components at every harmonic of 200 Hz. The filter allows only the carrier and components at ±200 Hz to pass. The answer is **C**.

Example 3.8

Given the periodic signal

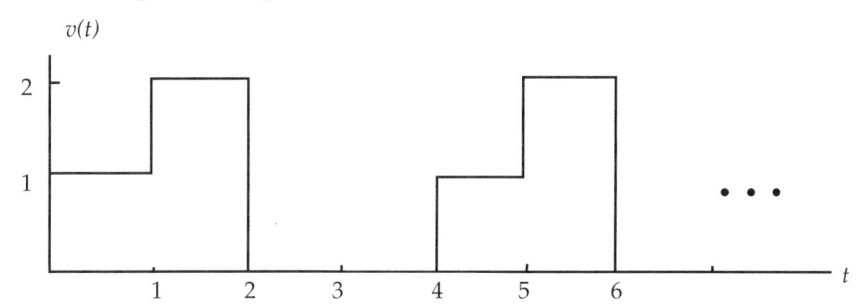

The exponential Fourier series for this signal is:

(A) $v(t) = \dfrac{1}{4} \sum\limits_{n} \left(\dfrac{\sin n\pi f_0}{n\pi f_0} \right) e^{-j\pi n f_0} e^{j2\pi n f_0 t}$

(B) $v(t) = \dfrac{1}{2} \sum\limits_{n} \left(\dfrac{\sin n\pi f_0}{n\pi f_0} \right) e^{-j3\pi n f_0} e^{j2\pi n f_0 t}$

(C) $v(t) = \dfrac{1}{2} \sum\limits_{n} \left(\dfrac{\sin n\pi f_0}{n\pi f_0} \right) e^{-j\pi n f_0} \left(\dfrac{1}{2} + e^{-2j\pi n f_0} \right) e^{j2\pi n f_0 t}$

(D) None of the above

Solution:

Using $\sum\limits_{n} c(n) e^{j 2 \pi n b f_0 t}$ calculate $c(n)$ or use transform techniques. The answer is **C**.

4. Solid State Devices

by Bong Ho

The material and equations needed to work problems involving Solid State devices are found on pages 99–103 of the NCEES Reference Handbook, 3rd ed. It will also be necessary to use equations found in the Electric Circuits section on pages 68–72. The several examples that follow illustrate the devices presented in the NCEES Reference Handbook.

Example 4.1 Semiconductor Material

A bar of intrinsic (pure) germanium is 2 cm long and has a cross-section area of 0.01 cm². A 10-volt battery is connected across it at room temperature.

Properties of germanium:

$$n_i = 2.4 \times 10^{13} \, \text{cm}^{-3}$$
$$\mu_n = 3900 \, \text{cm}^2/\text{V} \cdot \text{sec}$$
$$\mu_p = 1900 \, \text{cm}^2/\text{V} \cdot \text{sec}$$

1. What is the conductivity of the semiconductor material?
    (A) $3.5 \times 10^{-2} (\Omega \cdot \text{cm})^{-1}$
    (B) $2.2 \times 10^{-2} (\Omega \cdot \text{cm})^{-1}$
    (C) $4.7 \times 10^{-2} (\Omega \cdot \text{cm})^{-1}$
    (D) $1.1 \times 10^{-3} (\Omega \cdot \text{cm})^{-1}$

2. What is the current in the circuit?
    (A) 1.1×10^{-3} amp
    (B) 2×10^{-3} amp
    (C) 1.3×10^{-2} amp
    (D) 7.2×10^{-4} amp

3. How long does it take for an electron to drift the length of the bar?
   - (A) 0.1 sec
   - (B) 0.1 msec
   - (C) 0.1 μsec
   - (D) 10 μsec

4. How much power is dissipated in the bar?
   - (A) 1.1 W
   - (B) 2.1 W
   - (C) 0.11 W
   - (D) 0.011 W

Solutions:

1. **B** For an intrinsic semiconductor: $n_i = n = p$

$$\sigma = q(n\mu_n + p\mu_p)$$
$$= qn_i(\mu_n + \mu_p)$$
$$= 1.6 \times 10^{-19} \times 2.4 \times 10^{13}(3900 + 1900)$$
$$= 2.2 \times 10^{-2} (\Omega \cdot cm)^{-1}$$

2. **A** Electric field intensity:

$$E = \frac{V}{l} = \frac{10}{2} = 5 \frac{V}{cm}$$

   Current density:

$$J = \sigma E = 2.2 \times 10^{-2} \times 5 = 1.1 \times 10^{-1} \text{ A/cm}^2$$

   Current:

$$I = JA = 1.1 \times 10^{-1} \times 0.01 = 1.1 \times 10^{-3} \text{ A}$$

3. **B** Electron velocity:

$$v = \mu_n E = 3900 \times 5 = 1.95 \times 10^4 \text{ cm/sec}$$

   The time is then

$$t = \frac{l}{v} = \frac{2}{1.95 \times 10^4} = 0.1 \times 10^{-3} \text{ sec} = 0.1 \text{ msec}$$

4. **D** Power:

$$P = VI = 10 \times 1.1 \times 10^{-3} = 0.011 \text{ W}$$

Example 4.2 P-N Diode

A silicon *P-N* diode has a cross-section area of 10⁻³ cm². Both *P* and *N* sides are 0.1 cm long. At room temperature, the properties of Si are:

$$n_i = 1.5 \times 10^{10} \text{ cm}^{-3}, \quad \mu_n = 1.5 \times 10^3 \text{ cm}^2/\text{V} \cdot \text{sec}, \quad \mu_p = 6 \times 10^2 \text{ cm}^2/\text{V} \cdot \text{sec}$$

$$\rho_p = 10^{-3} \, \Omega \cdot \text{cm}, \quad \rho_N = 10^{-1} \, \Omega \cdot \text{cm}$$

1. What is the majority carrier concentration on the *P* side?
   - (A) 1×10^{17} cm⁻³
   - (B) 1×10^{18} cm⁻³
   - (C) 1×10^{19} cm⁻³
   - (D) 1×10^{20} cm⁻³

2. What is the minority carrier concentration on the *P* side?
   - (A) 22 cm⁻³
   - (B) 2.2×10^3 cm⁻³
   - (C) 2.2×10^4 cm⁻³
   - (D) 2.2×10^5 cm⁻³

3. What is the built-in potential of the diode?
   - (A) 0.09 V
   - (B) 0.45 V
   - (C) 0.7 V
   - (D) 0.9 V

4. Which side is at a higher (more positive) potential?
   - (A) N side
   - (B) P side
   - (C) Both sides are at the same potential
   - (D) Insufficient data

5. If the width of the junction is 0.1 μm $(1 \times 10^{-7}$ meter$)$, what will be the order of magnitude of the electric field intensity *E* inside the junction?
   - (A) 10³ V/m
   - (B) 10⁴ V/m
   - (C) 10⁵ V/m
   - (D) 10⁶ V/m

6. If the saturation current of the diode is $I_s = 4.13 \times 10^{-15}$ amp, what will be the diode current if 0.5 V is put across the diode?
   - (A) 1×10^{-6} amp
   - (B) 1×10^{-7} amp
   - (C) 1×10^{-8} amp
   - (D) 1×10^{-9} amp

Solutions:

1. **C** $\rho_p = \dfrac{1}{\sigma_p} = \dfrac{1}{p_p q \mu_p}$

   $\therefore p_p = \dfrac{1}{\rho_p q \mu_p} = \dfrac{1}{10^{-3} \times 1.6 \times 10^{-19} \times 6 \times 10^2} = 1 \times 10^{19} \text{ cm}^{-3} = N_a$

2. **A** $n_p = \dfrac{n_i^2}{p_p} = \dfrac{(1.5 \times 10^{10})^2}{0.1 \times 10^{20}} = 22 \text{ cm}^{-3}$

3. **D** $V_0 = \dfrac{kT}{q} \ln \dfrac{N_a N_d}{n_i^2} = \dfrac{1.38 \times 10^{-23}}{1.6 \times 10^{-19}} \ln \dfrac{4.17 \times 10^{16} \times 1 \times 10^{19}}{2.25 \times 10^{20}} = 0.9 \text{ V}$

   where we used

   $N_d = n_n = \dfrac{1}{\rho_N q \mu_n} = \dfrac{1}{10^{-1} \times 1.6 \times 10^{-19} \times 1.5 \times 10^3} = 4.17 \times 10^{16} \text{ cm}^{-3}$

4. **A** Donor ions are positive on the N side.

5. **D** $E = \dfrac{V_0}{d} = \dfrac{0.9}{0.1 \times 10^{-6}} = 9 \times 10^6 \text{ V/m}$

6. **A** For S_i, $n = 1$:

   $i_D = I_s \left[e^{\frac{V_D}{nV_T}} - 1 \right]$

   $= 4.13 \times 10^{-15} \left[e^{\frac{0.5}{0.026}} - 1 \right] = 1 \times 10^{-6} \text{ amp}$

Example 4.3 NPN Bipolar Junction Transistor BJT

A common-emitter BJT amplifier circuit is given below.

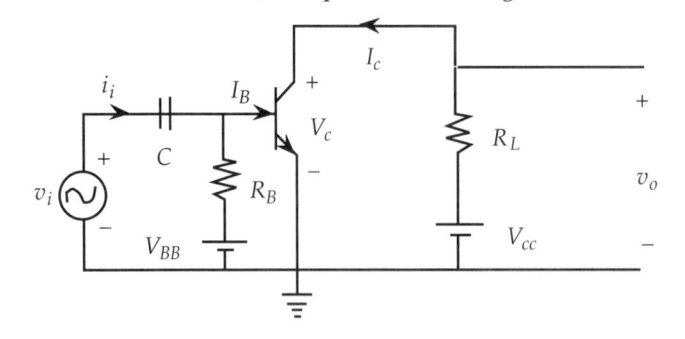

$V_{CC} = 12 \text{ V}$
$V_{BB} = 3 \text{ V}$
$R_B = 3.75 \text{ k}\Omega$
$R_L = 150 \text{ }\Omega$
$\beta = 90$
$r_\pi = 1 \text{ k}\Omega$
$C = \text{very large}$

1. Use the DC equivalent circuit to find the base current I_B.
   (A) 0.3 mA
   (B) 0.8 mA
   (C) 0.4 mA
   (D) 0.61 mA

2. Use the DC equivalent circuit to find the collector voltage V_C.

   (A) 3.76 V
   (B) 6 V
   (C) 4.21 V
   (D) 1.52 V

3. With an input signal $v_i = 0.1 \sin \omega t$ volt, use the low-frequency small signal equivalent circuit to find the voltage gain $A_V = \dfrac{V_o}{V_i}$.

   (A) −6.6
   (B) 15.2
   (C) −13.5
   (D) −37.8

4. Find the current gain $A_I = \dfrac{i_c}{i_i}$ for Problem 3.

   (A) 71
   (B) 52
   (C) −83
   (D) −31

Solutions:

1. **D** $I_B = \dfrac{V_{BB} - 0.7}{R_B} = \dfrac{3 - 0.7}{3.75 \times 10^3} = 0.61$ mA

2. **A** $I_C = \beta I_B = 90 \times 0.61 \times 10^{-3} = 54.9$ mA

   $V_C = V_{CC} - I_C R_L = 12 - 54.9 \times 10^{-3} \times 150 = 12 - 8.235 = 3.76$ V

3. **C** The AC equivalent circuit is

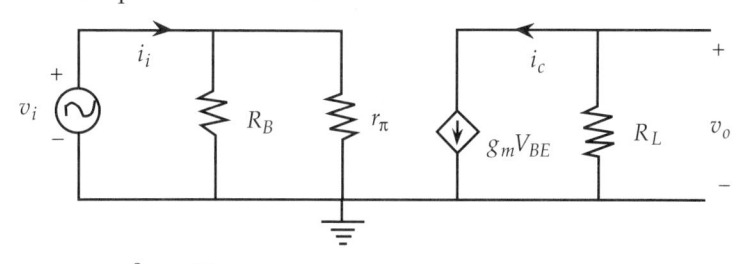

   $g_m = \dfrac{\beta}{r_\pi} = \dfrac{90}{10^3} = 9 \times 10^{-2}$ A/V

   $V_{BE} = v_i$

   $v_o = -g_m V_{BE} R_L = -9 \times 10^{-2} \times 0.1 \times 150 = -1.35 \sin \omega t$ volt

   Voltage Gain: $A_V = \dfrac{v_o}{v_i} = \dfrac{-1.35 \sin \omega t}{0.1 \sin \omega t} = -13.5$

4. **A** Refer to the AC equivalent above:

$$i_i = \frac{v_i}{R_B r_\pi / (R_B + r_\pi)} = \frac{0.1 \sin \omega t}{0.79 \times 10^3} = 0.126 \sin \omega t \text{ mA}$$

$$i_c = g_m V_{BE} = g_m v_i = 9 \times 10^{-2} \times 0.1 \sin \omega t = 9 \sin \omega t \text{ mA}$$

Current Gain: $A_I = \dfrac{i_c}{i_i} = \dfrac{9 \times 10^{-3} \sin \omega t}{0.126 \times 10^{-3} \sin \omega t} = 71.4$

Example 4.4 N-Channel Field-Effect Transistor FET

Given the FET amplifier circuit below with

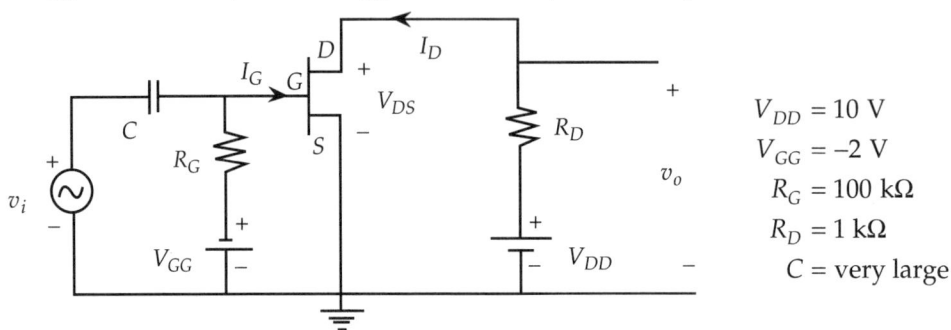

$I_{Dss} = 10$ mA, $V_P = -4$ V, $g_m = 2.5 \times 10^{-3}$ A/V and $r_d = 100$ kΩ

$V_{DD} = 10$ V
$V_{GG} = -2$ V
$R_G = 100$ kΩ
$R_D = 1$ kΩ
C = very large

The FET is operating in the saturation region.

1. What is the gate current I_G?
   (A) −0.02 mA
   (B) 0.02 mA
   (C) 0.013 mA
   (D) 0.0 mA

2. What is the drain voltage V_{DS}?
   (A) 7.5 V
   (B) 5 V
   (C) 3.5 V
   (D) 2.5 V

3. What is the voltage gain of the amplifier? (Using the AC equivalent circuit given.)
   (A) 5
   (B) −2.5
   (C) 2.5
   (D) −5

Solutions:

1. **D** Gate is open circuited by either reverse biasing the gate P-N junction in the case of JFET or by silicon dioxide layer in the MOSFET.

2. **A** Using the drain current expression in the saturation region,

$$i_D = I_{DSS}\left(1 - \frac{v_{GS}}{V_p}\right)^2$$

$$= 10 \times 10^{-3}\left(1 - \frac{-2}{-4}\right)^2 = 2.5 \times 10^{-3} \text{ amp}$$

Then,

$$v_{DS} = V_{DD} - I_D R_D$$

$$= 10 - 2.5 \times 10^{-3} \times 10^3 = 7.5 \text{ V}$$

3. **B** The AC equivalent circuit is

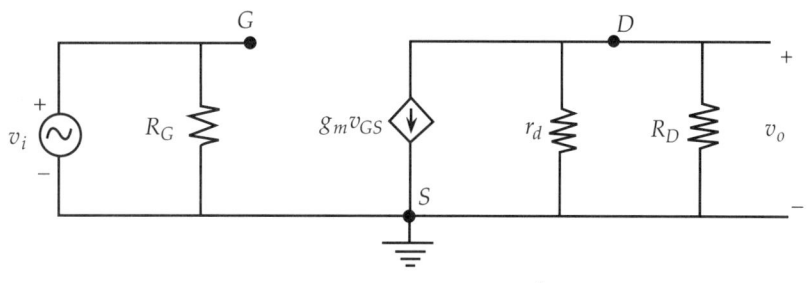

$$g_m = \frac{2\sqrt{I_{DSS} I_D}}{|V_p|} = \frac{2\left[10 \times 10^{-3} \times 2.5 \times 10^{-3}\right]^{\frac{1}{2}}}{4} = 2.5 \times 10^{-3} \text{ A/V}$$

$$v_o \approx -g_m v_{GS} R_D = -g_m R_D v_i \quad (v_{GS} = v_i)$$

Voltage Gain: $\quad A_v = \dfrac{v_o}{v_i} = -g_m R_D = -2.5 \times 10^{-3} \times 10^3 = -2.5$

5. Computer Engineering

by R. Lal Tummala

Many of the questions relating to Computer Engineering can be answered using the equations and material found on page 104 of the NCEES Handbook, 3rd ed. So make sure you are familiar with that material. This material is reviewed in the following sections. Additional discussion is provided whenever it is appropriate to complement the material in the Handbook.

5.1 Number Systems and Codes

This section deals with converting numbers from one system to another.

Example 5.1

What is the decimal equivalent of octal number 555 ?
- (A) 556
- (B) 458
- (C) 365
- (D) 326

Solution:

The base for the octal number is 8. Decimal equivalent of octal number 555 = $5 \times 8^2 + 5 \times 8^1 + 5 \times 8^0 = 320 + 40 + 5 = 365$. The answer is **C**.

Example 5.2

What is the decimal equivalent of the hexadecimal number FE?
- (A) 290
- (B) 255
- (C) 128
- (D) 254

Solution:

The base for the hexadecimal number is 16. The decimal equivalent of
FE = F x 16^1 + E x 16^0 = 15 x 16 + 14 = 240 + 14 = 254. The answer is **D**.

Example 5.3

What is the decimal equivalent of the following binary coded decimal number:
0001 1001 0110 0111?

(A) 1965
(B) 4058
(C) 1967
(D) 850

Solution:

Divide the number into groups of four starting from the right.

    0001   1001   0110   0111
      1   9   6   7

The answer is C.

Example 5.4

The hexadecimal equivalent of the 16 bit binary number 1110 1011 1000 0011 is:
(A) EFE3
(B) EEFF
(C) EB85
(D) EB83

Solution:

Group the binary digits into groups of four from right and replace each group by the equivalent hexadecimal digit. The answer is **D**.

Decimal Equivalent

Decimal equivalent of a number is obtained by multiplying with the positive powers of the base for the digits to the left of the radix point and negative powers of the base to the radix point.

Example 5.5

What is the decimal equivalent of binary number 10011.101 ?
(A) 13.4
(B) 13.5
(C) 19.625
(D) 19.5

Solution:

The base for binary numbers is 2

$(1 \times 2^4 + 0 \times 2^3 + 0 \times 2^2 + 1 \times 2^1 + 1 \times 2^0) . (1 \times 2^{-1} + 0 \times 2^{-2} + 1 \times 2^{-3}) = 19.625$

The answer is **C**.

Conversion of Decimal Numbers

Conversion of decimal numbers to binary, octal or hexadecimal numbers is accomplished by "repeated division" by the base. The process is stopped when the quotient becomes zero. This is illustrated by the following example.

Example 5.6

Convert a decimal number 29 to a binary number.

(A) 1 1 1 1 0

(B) 1 0 1 1 1

(C) 1 1 0 1 1

(D) 1 1 1 0 1

Solution:

29/2	=	14 + 1	Remainder	=	1
14/2	=	7 + 0	Remainder	=	0
7/2	=	3 + 1	Remainder	=	1
3/2	=	1 + 1	Remainder	=	1
1/2	=	0 + 1	Remainder	=	1

The process stops because the quotient is zero. The remainder obtained in each step is the desired digit for the binary number in the reverse order shown by the arrow.

The corresponding binary number for decimal 29 is: 11101. The answer is **D**.

Alternate Method of Conversion

An alternate method for large numbers is to convert a decimal number into hexadecimal (octal) and replace each hexadecimal (octal) number by the corresponding binary number.

Example 5.7

Convert a decimal number 29 to binary number by first converting to an octal number.

Solution:

29/8	=	3 + 5	Remainder	=	5
3/8	=	0 + 3	Remainder	=	3

The octal number : 35 and the binary number: 011101. Since the left most zero has no meaning, it is usual practice to leave out the left-most zeros. The answer = 11101.

Example 5.8

The binary number for a decimal number 982 is :
(A) 11101111
(B) 11101011
(C) 1112111
(D) 1111010110

Solution:

$$982/16 = 61 + 6 \quad \text{Remainder} = 6$$
$$61/16 = 3 + 13 \quad \text{Remainder} = 13 = D \text{ in hexadecimal number system}$$
$$3/16 = 0 + 3 \quad \text{Remainder} = 3$$

The hexadecimal number = 3D6

Replace each hexadecimal digit by four bit binary equivalent = 0011 1101 0110. Leave out the two leftmost zeroes and the answer is 1111010110. The answer is **D**.

5.2 Binary Addition and Subtraction

The binary addition and subtraction is performed exactly like the decimal arithmetic. It is carried from right to left. The difference is that in binary arithmetic, we only have 1 and 0 and so the carry occurs whenever the sum in each column exceeds 2. Binary subtraction can be done similarly using the concept of *borrowing*, familiar in decimal arithmetic. In practice, however, subtraction is accomplished using 2's complement arithmetic. Negative numbers are represented as 2's complement of the corresponding positive number. In the computer the 2's complement of a binary number is obtained by complementing (replacing 1 by 0 and vice versa) and adding 1 to the result. So, to obtain $A - B$, we write it as: $A+(-B)$ where $-B$ represents the 2's complement of B.

Example 5.9

Given: $A = 00111$, $B = 00101$. Find $A + B$.

Solution:

A	0	0	1	1	1	
B	0	0	1	0	1	
			1	1		← Carry digit
A+ B =	0	1	1	0	0	

Example 5.10

Two binary numbers, $A = 00111$, $B = 00101$ are given.

1. Find $A + B$.
   - (A) 10000
   - (B) 10101
   - (C) 01100
   - (D) 00212

2. Find the 2's complement of B.
   - (A) 11000
   - (B) 11011
   - (C) 00112
   - (D) 00111

3. Find $A - B$.
   - (A) 00010
   - (B) 11001
   - (C) 00110
   - (D) 00202

Solution:

1. **C.** (See Example 5.9)

2. **B.** Given $B = 00101$
   Complement each bit = 11010
   Add 1 to the result   +   1

   2's complement of B = 11011 = $-B$

3. **A.** The subtraction is done by adding : $A + (-B)$ and ignoring the carry bit.

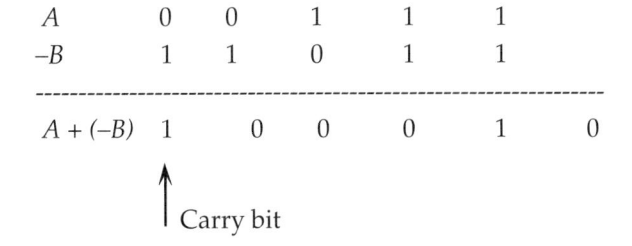

A	0	0	1	1	1
$-B$	1	1	0	1	1
$A + (-B)$	1	0	0	0	1	0

   ↑ Carry bit

5.3 Logic Operations and Boolean Algebra

Boolean logic deals with zeros (0) and ones (1). Three basic logic operations, "AND (.)", "OR (+)" and "Exclusive-OR ($\oplus$)" along with their symbols are given in the Handbook. Boolean algebra deals with the algebra of binary variables, i.e., variables that can take only two values: "0" (False) and "1" (True).

Logic circuits are divided into two classes: *Combinational Circuits* and *Sequential Circuits*. A combinational circuit consists of combinations of logic gates like AND, OR, Exclusive-OR, etc., whose output depends only on the present inputs. On the other hand, sequential circuits consist of flip-flops and logic gates, and their output depend not only on the present inputs but also on the previous outputs.

Designing of combinational logic circuits starts with the mapping of the problem into a *truth table*. A truth table defines the output for all possible combinations of the input variables. The output is either "1", "0" or don't cares, denoted by "X". Don't cares represent the outputs due to input combinations that cannot occur. This information from the truth table is used to define Boolean equations. The structure of the Boolean equation is of the form:

$$X = F(A, B, C..)$$

where Boolean variable X represents the output and Boolean variables $A, B, C..$ represent the inputs to the logic circuit. For example, the following equation:

$$X = AB + AC + CDE$$

implies that X is *true* if A and B are true, OR A and C are true OR, C and D and E are true.

Boolean algebra is used to simplify these functions. Finally, logic gates are used to implement the simplified function. Some important theorems useful in simplification are given below.

In the following theorems, "+" represents an OR operation and "." represents an AND operation.

(1 = TRUE, 0 = FALSE.)

1. $A + 1 = 1$
2. $A + 0 = A$
3. $A \cdot 1 = A$
4. $A \cdot 0 = 0$
5. $A + A = A$
6. $A \cdot A = A$
7. $A + \overline{A} = 1$
8. $A \cdot \overline{A} = 0$
9. $A(B + C) = AB + AC$
10. $A + AB = A$
11. $A(A + B) = A$
12. DeMorgan's theorems: $\overline{AB} = \overline{A} + \overline{B}$; $\overline{A + B} = \overline{A}\,\overline{B}$

The above concepts are illustrated with the following example.

Example 5.11

Given the following truth table,
  a) Write the logic function
  c) Simplify the function
  d) Implement the function using logic gates.

Truth Table

A	B	C	F
0	0	0	1
0	0	1	1
0	1	0	0
0	1	1	1
1	0	0	1
1	0	1	0
1	1	0	0
1	1	1	0

Solution:

a) $F = \overset{1}{\overline{A}\,\overline{B}\,\overline{C}} + \overset{2}{\overline{A}\,\overline{B}C} + \overset{3}{\overline{A}BC} + \overset{4}{A\overline{B}\,\overline{C}}$

b) Combining terms **1** and **4** & **2** and **3**, we get

$$F = (A + \overline{A})\overline{B}\,\overline{C} + \overline{A}C(B + \overline{B})$$

Using Theorem 7 above, we have

$$F = \overline{B}\,\overline{C} + \overline{A}C$$

c) There are many ways to implement this function. Two different implementations are shown below.

  i) AND-OR-NOT implementation:

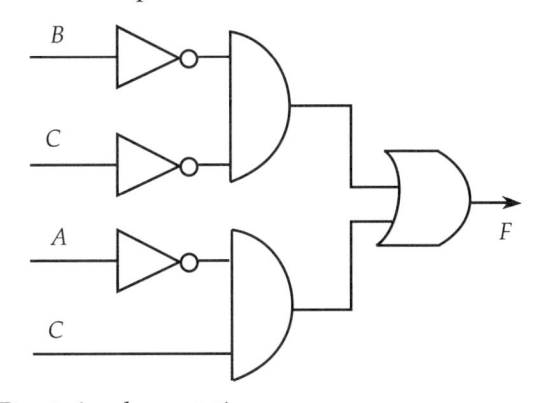

  ii) NAND gate implementation:

  $$F = \overline{B}\,\overline{C} + \overline{A}C$$

Applying DeMorgan's theorem:

$$\bar{\bar{F}} = \overline{\overline{BC} + \overline{AC}} = \overline{\overline{BC}} \cdot \overline{\overline{AC}}$$

Example 5.12

Given the following digital circuit:

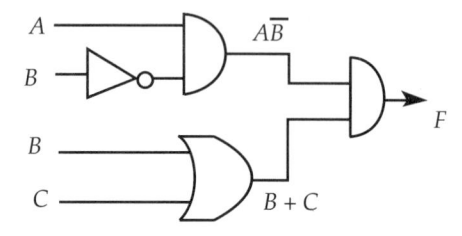

1. The equation for the function F after simplification is:

   (A) $A\,\bar{B}\,C$

   (B) $A\,B\,C$

   (C) $A\,\bar{B} + \bar{A}\,B$

   (D) $B\,\bar{C}$

2. The equation for $\bar{F}$ is:

   (A) $A\,\bar{B}\,C$

   (B) $\bar{A} + B + \bar{C}$

   (C) $A\,B + \bar{C}$

   (D) $A + B + \bar{C}$

Solutions:

1. **A.** $F = A\bar{B}\,(B+C) = A\bar{B}B + A\bar{B}C = 0 + A\bar{B}C = A\bar{B}C$

2. **B.** Using DeMorgan's theorem:

   $$\bar{F} = \overline{A\bar{B}C} = \bar{A} + B + \bar{C}$$

Example 5.13

The truth table of a digital logic circuit is given below.

A	B	F
0	0	1
0	1	1
1	0	1
1	1	0

The expression for the simplified logic function F is:

(A) $F = A + \overline{B}$

(B) $F = \overline{A}\,\overline{B}$

(C) $F = A + B$

(D) $F = \overline{AB}$

Solution:

$$F = \overline{A}\,\overline{B} + \overline{A}B + A\overline{B}$$
$$= \overline{A}\,\overline{B} + \overline{A}\,\overline{B} + \overline{A}B + A\overline{B} = \overline{B}(A + \overline{A}) + \overline{A}(B + \overline{B}) = \overline{A} + \overline{B} = \overline{AB}$$

The answer is **D**.

5.4 Karnaugh Maps

A Karnaugh map (K-map) is another method used to simplify logic functions. Even though it can be used for any number of variables, it loses its usefulness beyond four variables. The K-maps for 2, 3 and 4 variables are shown below.

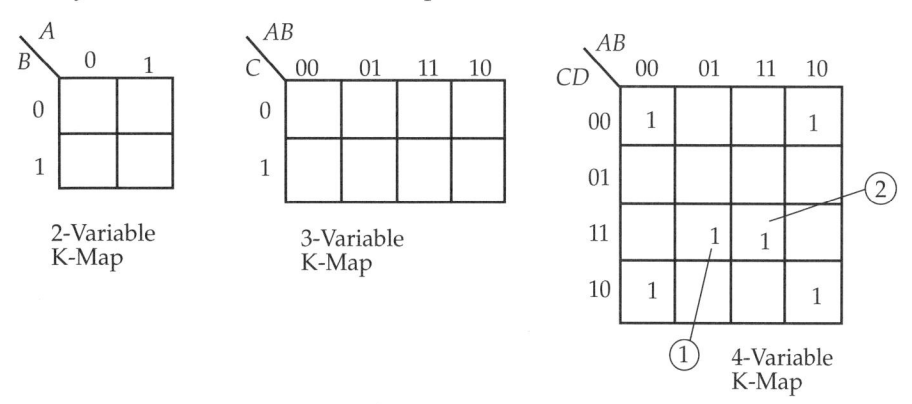

Each square of the map is identified by a certain combination of the variables. For example, the square marked "②" represents $AB\overline{C}D$. Notice, by design, adjacent squares vertically or horizontally differ by only one variable change. These adjacent squares can be combined to eliminate variables. For example, in the 4-variable map, **1** and **2** can be combined to eliminate the variable A, resulting in a term $BC\overline{D}$. Similarly four terms represented by "1"s in the corners of the 4-variable map can be combined to produce a term, $\overline{B}\,\overline{D}$. Entries in each square represent the output values corresponding to each row in the truth table. A map is created for each output variable. The plotted function in the K-map is simplified by encircling the largest adjacent squares of 1's. Six simple rules are given below. These rules are applied to each square containing the 1's in the K-map.

1. All encirclements should contain 2^N 1's, where N is an integer.

2. Identify the adjacent squares containing "1"s that can be combined with a single other box in only one way. Encircle such two square combinations. Skip any square that can be combined more than one way.

3. Identify the squares that can be combined with three other squares in only one way. If all four squares so involved are not already covered in groupings of two, encircle these four squares. Skip any square that can be combined in a group of four in more than one way.

4. Repeat the procedure for groups of eight, etc.

5. After the above procedure, if there still remain some uncovered squares, they can be combined with each other or with other squares already used. If no square can be combined with the others, they appear in the final result without any simplification.

6. Don't cares may be used to simplify functions in the K-map, but are not included in the final logic equation.

Example 5.14

The following function is obtained from the truth table.

$$F = AB\overline{C} + A\overline{B}\,\overline{C} + \overline{A}\,\overline{B}\,\overline{C}$$
$$110 \quad\ \ 100 \quad\ \ 000$$

Plot the function on 3 variable K-map and simplify the function.

Solution:

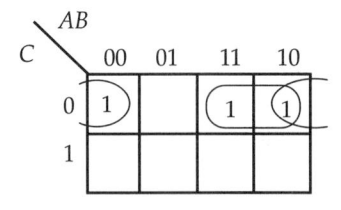

C \ AB	00	01	11	10
0	1		1	1
1				

The simplified function is:

$$F = A\overline{C} + \overline{B}\,\overline{C}$$

Example 5.15

Simplify the following function given on 4-variable K-map.

The simplified function is given by:

(A) $A\bar{D}B + C\bar{D}$

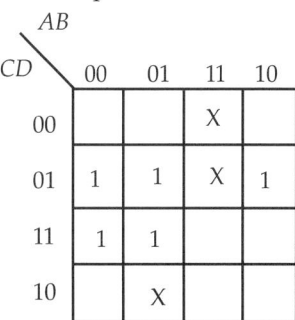

(B) $\bar{A}D + \bar{C}D$

(C) $AD + \bar{C}D$

(D) $A\bar{B}D + C\bar{D}$

CD \ AB	00	01	11	10
00			X	
01	1	1	X	1
11	1	1		
10		X		

Solution:

The answer is **B**.

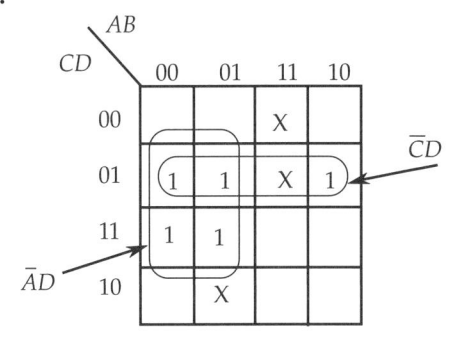

CD \ AB	00	01	11	10
00			X	
01	1	1	X	1
11	1	1		
10		X		

$\bar{C}D$ groups the 01 row; $\bar{A}D$ groups the left columns rows 01 and 11.

5.5 Flip-Flops

As mentioned earlier, *sequential circuits remember the past. The output of a sequential circuit depends not only on the present input but also on the previous outputs (caused by the past inputs).* Flip-flops along with the logic gates are used for this purpose. Three basic flip-flops are described in the NCEES Handbook. They are *RS* flip-flop, *JK* flip-flop and *D* flip-flop. The flip-flop output is synchronized with a clock (CLK) signal. Q_n represents the output before the clock pulse is applied. Q_{n+1} represents the output of the flip-flop after the clock signal is applied. The output Q_{n+1} depends on the signal present on the input terminals (e.g., J and K for JK flip-flop) and the previous output, Q_n. The truth tables are given in the NCEES Handbook on page 104.

The Composite Flip-Flop transition tables given in the NCEES Handbook on page 99 are used to determine the corresponding inputs required on the flip-flop inputs to accomplish the change in the output. Since this change also depends on the previous output, it has to be taken into consideration in determining the inputs. For example, if we want to change the output of JK flip-flop from $Q_n = 1$ to $Q_{n+1} = 0$, then we should set $K = 1$ and $J = x$, where "x" represents don't care, i.e., J could be either "0" or "1" before the clock signal is applied.

Example 5.16

An Exclusive-OR logic gate is added to a *JK* flip-flop as shown below.

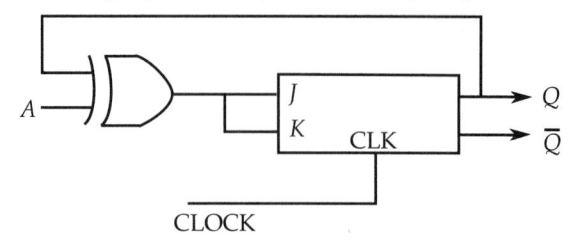

The new circuit represents a :

(A) *JK* flip-flop

(B) *D* FF

(C) *SR* FF

(D) None of the above

Solution:

A	Q_n	Q_{n+1}
0	0	0
0	1	0
1	0	1
1	1	1

Type *D* flip-flop. The answer is **B**.

Example 5.17

A sequential circuit is shown below. In the circuit, Q^1 is the most significant bit and Q^0 is the least significant bit, i.e., state 10 means $Q^1 = 1, Q^0 = 0$.

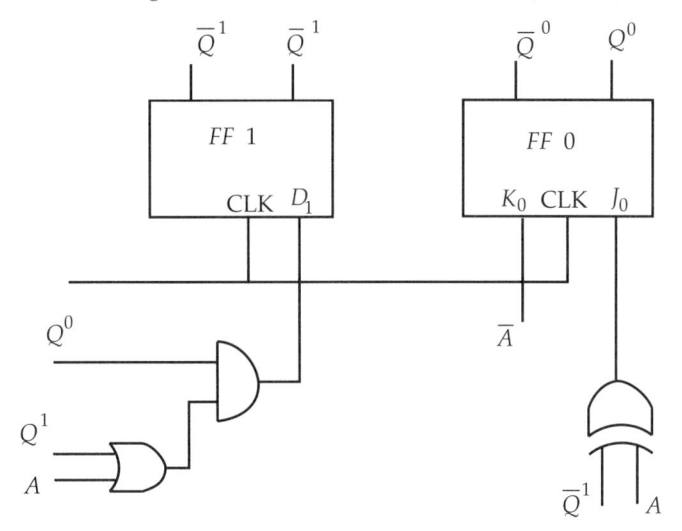

1. If the sequential circuit is in 00 state, what is the next state of the circuit if $A = 0$?
   (A) 01
   (B) 00
   (C) 10
   (D) 11

2. If the circuit's initial state is 01, what is the next state if $A = 1$?
   (A) 00
   (B) 01
   (C) 11
   (D) 01

3. If the sequential circuit is in 00 state, what input should we apply to the JK terminals and D terminals of the flip-flops to change it to 01?
   (A) $J = K = 0, D = 0$
   (B) $J = 1, K = X, D = 0$
   (C) $J = 1, K = 1, D = 1$
   (D) None of the above

4. If the sequential circuit is in 00 state, what should be the value of A to change it to 01?
   (A) $A = 0$
   (B) $A = 1$
   (C) $A = 01$
   (D) $A =$ don't care

5. The output of the circuit is defined by the following logic circuit. What is the output equation?

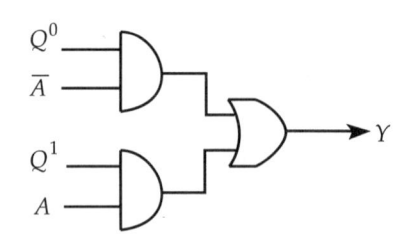

(A) $Y = Q^1 A$
(B) $Y = Q^0 \overline{A} + Q^1 A$
(C) $Y = Q^0 A + B$
(D) $Y = Q^1 + Q^0$

Solution:

The answers are easily obtained by writing the following logic equations from the circuit.

$$J_0 = A \oplus \overline{Q}^1, K_0 = \overline{A}$$
$$D_1 = (A + Q^1) \cdot Q^0$$

1. A $J_0 = 1, K_0 = 1, D_1 = 0 \rightarrow Q^0_{n+1} = 1, Q^1_{n+1} = 0$
2. C $J_0 = K_0 = 0, D_1 = 1$
3. B
4. A
5. B

6. Control Systems

by R. Lal Tummala

Control systems operate either in *open-loop* or *closed-loop*. In open-loop systems, the output has no effect on the input. On the other hand, closed-loop systems continuously sense the output and make appropriate adjustments to the input to keep the output at the desired level. An example is shown in Fig. 6.1 where the speed of the motor is controlled by an input voltage. Figure 6.1a represents an open-loop control system where there is no mechanism to adjust input as the output deviates from the desired level, (example, to load variations). On the other hand, the closed-loop control system shown in Fig. 6.1b automatically senses the changes of the output and sends a corrective voltage to return the motor speed to the desired level.

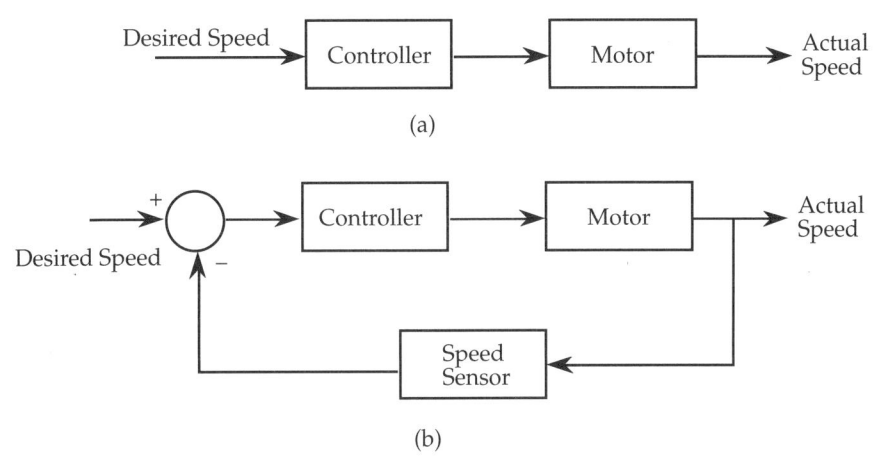

Figure 6.1 a) Open-loop system. b) Closed-loop system.

Analysis and design of *closed-loop control systems*, also called *feedback control systems* is based on well defined performance criteria, such as stability, sensitivity, accuracy, transient response, bandwidth, and noise rejection. Several approaches were developed in the past for this purpose. We will discuss these approaches in this chapter.

6.1 Transfer Functions

Majority of the analyses and designs of feedback systems are based on transfer functions. The transfer function is an input-output representation in the s-domain and is defined as: Laplace transform of the output divided by the Laplace transform of the input when all the initial conditions are assumed to be zero.

A system can contain many components and the overall transfer function of the system can be obtained by using the following information.

Series Connection

If two transfer functions are connected in series as shown in Fig. 6.2 , that is, if

$$Y_1(s) = G_1(s)R_1(s)$$
$$Y_2(s) = G_2(s)R_2(s) \qquad (6.1.1)$$

then

$$Y_2(s) = G_1(s)G_2(s)R_1(s) \qquad (6.1.2)$$

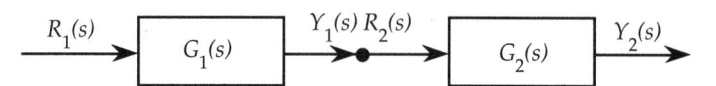

Figure 6.2 Series connection.

Parallel Connection

On the other hand, if the transfer functions are connected in parallel as shown in Fig. 6.3, then the overall transfer function is given by:

$$Y(s) = G_1(s)R(s) + G_2(S)R(s) = [G_1(s) + G_2(s)]R(s) \qquad (6.1.3)$$

or

$$\frac{Y(s)}{R(s)} = G_1(s) + G_2(s) \qquad (6.1.4)$$

Figure 6.3 Parallel connection.

Closed-Loop Transfer Function

The transfer function of a closed-loop system given in Fig. 6.4 is obtained as follows:

$$E(s) = R(s) - B(s) = R(s) - H(s)Y(s) \qquad (6.1.5)$$

and

$$Y(s) = G(s)E(s) \qquad (6.1.6)$$

Using the above information, we can get the closed-loop transfer function,

$$\frac{Y(s)}{R(s)} = \frac{G(s)}{1+G(s)H(s)} \qquad (6.1.7)$$

When $H(s) = 1$, the systems are called *unity feedback* systems.

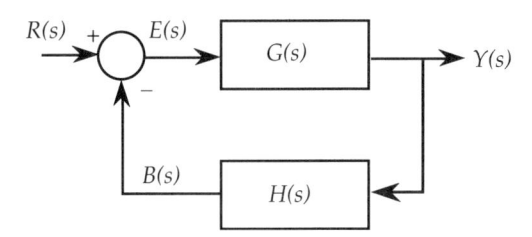

Figure 6.4 Closed-loop system.

Multiple Inputs

Since we are dealing with linear time invariant systems, we can obtain the output of a system for multiple inputs by using superposition principle. In this case, we determine the output due to individual inputs and sum them. The following example illustrates this method.

Example 6.1

Given the following system, find the output $Y(s)$.

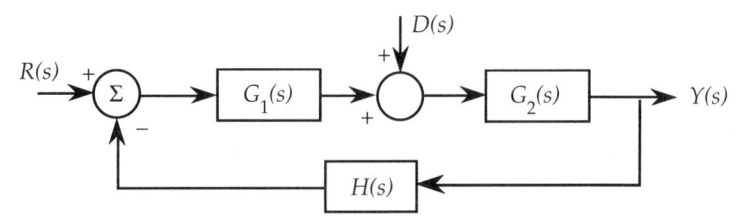

Set $R(s) = 0$ and find the output due to $D(s)$, using Eq.6.1.7:

$$Y_1(s) = \frac{G_2}{1+G_1G_2H}D(s)$$

Similarly, set $D(s) = 0$ and find the output due to $R(s)$:

$$Y_2(s) = \frac{G_1G_2}{1+G_1G_2H}R(s)$$

The total output is

$$Y(s) = Y_1(s) + Y_2(s)$$

6.2 Poles, Zeros and Stability

The poles and zeros of a transfer function $Y(s)/R(s)$ are given by the values of s where $Y(s)/R(s) = \infty$ and $Y(s)/R(s) = 0$, respectively. For example, if

$$\frac{Y(s)}{R(s)} = \frac{(s-2)(s+1)}{s(s+5)(s^2+2s+2)} \tag{6.2.1}$$

then the zeros are $s = 2, -1$ and poles are at $s = 0, -5, -1 \pm j1$. Thus, the poles and zeros can be real or complex. The plot of poles and zeros for the transfer function is shown in Fig. 6.5 where X is used represent poles and O is used to represent zeros.

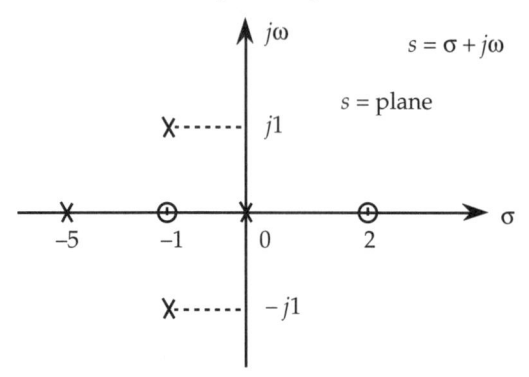

Figure 6.5 Pole-zero plot of the transfer function.

The time response of the system consists of two parts: *transient response* and *steady state response*. The steady state response of the system is the behavior of the system as t approaches infinity. The transient response, roughly speaking, is the intermediate behavior of the system between the initial state and the steady state. For example, if

$$\frac{Y(s)}{U(s)} = \frac{2s+1}{s^2+3s+2} \tag{6.2.2}$$

then the system has poles at $s = -1$, $s = -2$ and zero at $s = -0.5$ Now, if we apply a unit step input to the system, we have:

$$y(t) = \frac{1}{2} + e^{-t} - \frac{3}{2}e^{-2t} \tag{6.2.3}$$

We can see from Eq. 6.2.3 that the shape of the transient response is governed by the poles of the system. This is true of more complex systems as well and thus the design of many systems involve locating these poles in the s-plane to achieve desired transient response. The zeros influence the size of the coefficient that multiplies the transient components. It is also evident from Eq. 6.2.3, that if the poles have positive real parts, then the transient response grows with time. We call these *unstable systems*. Transient response of the *stable systems* go to zero as t approaches infinity. Thus we can determine the stability of the system from the locations of the poles in the s-plane. For a general closed-loop system, the closed-loop poles are given by solving the following equation:

$$1 + G(s)H(s) = 0 \tag{6.2.4}$$

Since the roots of this equation define the transient characteristics of the system, this is also called the *characteristic equation*. The roots of this equation are called *characteristic roots*.

Example 6.2

For a closed-loop feedback system shown
$$G_1(s) = \frac{10(s+5)}{s}$$

The input-output $G_2(s)$ of the process is given by the differential equation
$$y'' + 5y' + 6y = m(t)$$

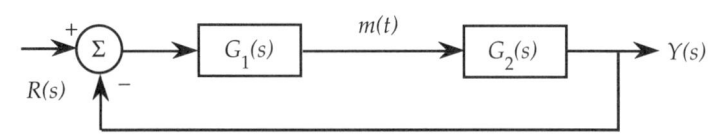

a) Find the closed-loop transfer function.
b) Find the poles and zeros of the closed-loop transfer function.
c) Find the characteristic equation.
d) Find the characteristic roots.

Solution: a) Taking Laplace transforms and setting initial conditions, the transfer function is
$$G_2(s) = \frac{1}{(s+2)(s+3)}$$

Using series connection, we get
$$G(s) = G_1(s)G_2(s)$$

Using the Eq. 6.1.7, with $G(s)$ given above and $H(s) = 1$, we get the total transfer function
$$\frac{Y(s)}{R(s)} = \frac{G(s)}{1 + G(s) \times 1} = \frac{10(s+5)}{s(s+2)(s+3) + 10(s+5)}$$

b) The poles of the closed-loop transfer function are : $-4.0814, -0.459 \pm j3.47$

The zeros of the closed-loop transfer function are : -5

c) The characteristic equation is : $s^3 + 5s^2 + 16s + 50 = 0$

d) The characteristic roots are : $-4.0814, -0.459 \pm j3.47$

Sensitivity

Define

$$T(s) = \frac{Y(s)}{R(s)} = \frac{G(s)}{1+G(s)H(s)} \quad (6.2.5)$$

for the feedback system shown in Fig. 6.4. Then the sensitivity for changes in $G(s)$ is defined as

$$S_G^T = \frac{\partial T}{\partial G}\frac{G}{T} = \frac{1}{1+GH} \quad (6.2.6)$$

The sensitivity to changes in $H(s)$ is given by

$$S_H^T = \frac{\partial T}{\partial H}\frac{H}{T} = \frac{-GH}{1+GH} \quad (6.2.7)$$

If we are interested in the sensitivity of a parameter, say α, within a transfer function $G(s)$, then the sensitivity of T with respect to α can be obtained by the chain rule:

$$S_\alpha^T = S_G^T S_\alpha^G \quad (6.2.8)$$

6.3 Time Domain Performance

First Order Systems

The input-output representation of a first order system (see Fig. 6.6) given by the transfer function

$$G(s) = \frac{Y(s)}{R(s)} = \frac{1}{s\tau + 1} \quad (6.3.1)$$

The unit step response of the system is given by

$$y(t) = 1 - e^{-t/\tau} \text{ for } t \geq 0 \quad (6.3.2)$$

Note that the output $y(t)$ reaches 66.2% of its final value at $t = \tau$ and τ is called the *time constant*. The first order system reaches 98% of its final value in about 4 time constants.

Second Order Systems

Traditionally, second order systems are used to specify the transient response. It was found that the dominant behavior of many higher order systems can be approximated by second order systems. The closed-loop transfer function of a general second order system is given by

$$\frac{Y(s)}{R(s)} = \frac{\omega_n^2}{s^2 + 2\xi\omega_n s + \omega_n^2} \quad (6.3.3)$$

where ω_n is the natural frequency of the system and ξ is called the damping ratio. The step response is given by

$$y(t) = 1 - \frac{e^{-\xi\omega_n t}}{\sqrt{1-\xi^2}} \sin\left(\omega_n\sqrt{1-\xi^2}\, t + \tan^{-1}\frac{\sqrt{1-\xi^2}}{\xi}\right) \quad (6.3.4)$$

The response $y(t)$ and the corresponding performance criteria are given in Fig. 6.7.

Maximum overshoot: Largest deviation of the output over the desired output during the transient state. This is usually expressed as a percentage of the output:

$$\text{overshoot} = 100 e^{-\pi\xi/\sqrt{1-\xi^2}} \quad (6.3.5)$$

Rise time: Time required for the output to rise from 10% of the final value to 90% of the final value. It measures the speed of the response:

$$t_r = \frac{0.8 + 2.5\xi}{\omega_n} \quad (6.3.6)$$

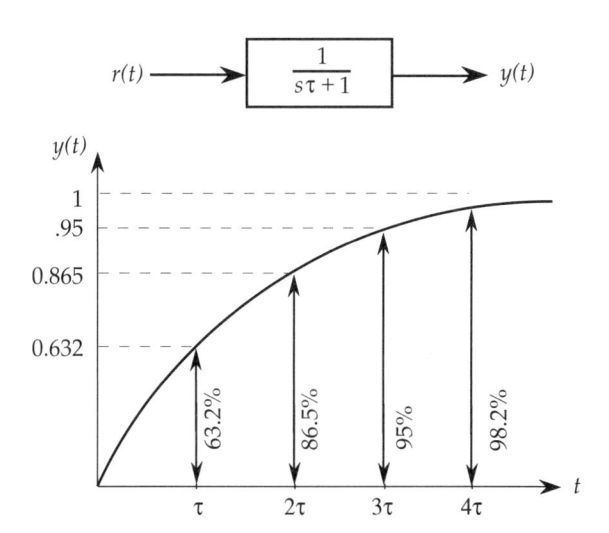

Figure 6.6 Step response of a first order system.

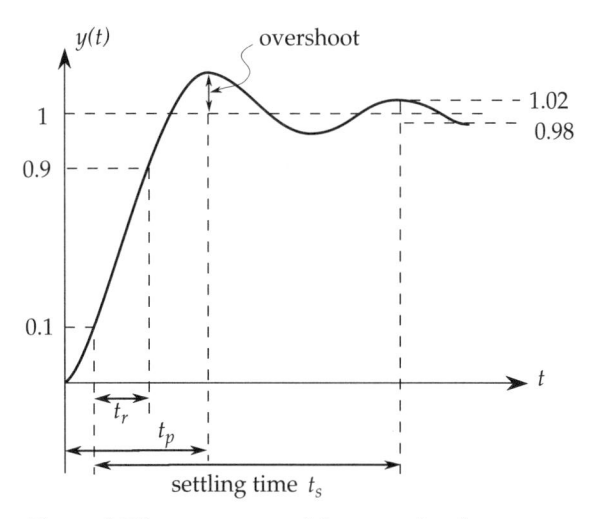

Figure 6.7 Step response of the second order system.

Peak time: Time it takes to reach the first peak as shown in Fig. 6.7. Peak time is also used in conjunction with rise time to define how fast the system responds to inputs:

$$t_p = \frac{\pi}{\omega_n\sqrt{1-\xi^2}} \qquad (6.3.7)$$

Settling time: Time it takes for the system to settle within certain percentage of the final output. Typical values used are 2% and 5%:

$$t_s = \frac{4}{\xi\omega_n} \quad \text{for } 2\% \qquad (6.3.8)$$

The above formulas are for $0 < \xi < 1$.

The performance measures are easy to obtain once the response is plotted. This can also be determined from the locations of the closed-loop poles (characteristic roots) in the *s*-plane as shown in Fig. 6.8. The characteristic equation is given by:

$$s^2 + 2\xi\omega_n s + \omega_n^2 = 0 \qquad (6.3.9)$$

The characteristic roots are

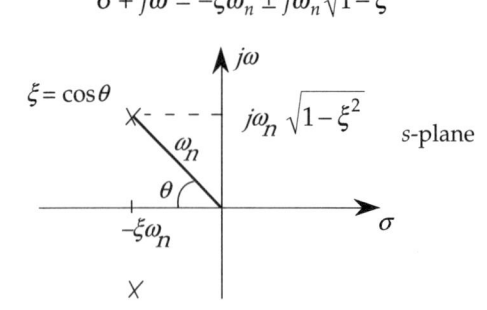

$$\sigma + j\omega = -\xi\omega_n \pm j\omega_n\sqrt{1-\xi^2} \qquad (6.3.10)$$

Figure 6.8 Characteristic roots of a second order system in the *s*-plane.

It can be seen from Eq. 6.3.4 that $\xi\omega_n$ controls the rate of rise or decay of the time response and is usually called the *damping factor*. The inverse $\xi\omega_n^{-1}$ is called the *time constant*. When the two characteristic roots are equal, the system is called critically damped. This occurs when $\xi = 1$. In this case, the damping factor is ω_n. Therefore, the parameter ξ can be regarded as the damping ratio which is the ratio between the actual damping factor and the damping factor when $\xi = 1$. When $\xi = 0$, the roots of the characteristic equation are $\pm j\omega_n$. Therefore ω_n is called the undamped natural frequency.

Higher-Order Systems

The relationships given above are only valid for pure second order systems. However, majority of the practical control systems are of higher order, and the determination of the transient response is determined by using the *dominant pole* concept. This is derived from the fact that the poles close to the imaginary axis in the left-half of the *s*-plane decay slowly compared to the poles that are far away. Using this fact, the higher order systems, in practice are approximated by lower order systems by keeping the poles near the imaginary axis (usually called *dominant poles*) and neglecting all the poles whose real parts are 5 to 10 times the real parts of the dominant poles for preliminary analysis and design. When this approximation is used, the damping ratio defined above is called the relative damping ratio.

Example 6.3

The following control system is given with the transfer function

$$G_p(s) = \frac{1}{s(s+a)}$$

It is required to determine the value of K and a such that the transient response to a step input satisfies the following:

– Percent overshoot is less than 5%
– Settling time (2%) less than 4 seconds

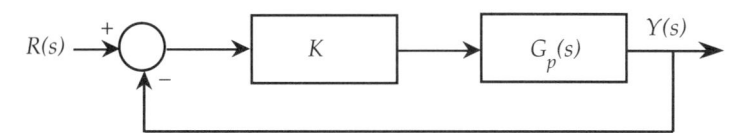

Solution: To satisfy the requirements, we have to determine the closed-loop poles (characteristic roots) of the system. The characteristic equation is given by

$$1 + \frac{K}{s(s+a)} = 0 \quad \text{or} \quad s^2 + as + K = 0$$

Comparing with the system given in Eq (6.3.9), we have

$$a = 2\xi\omega_n, \quad K = \omega_n^2$$

Percent overshoot less than 5% gives from Eq.6. 6.5,

$$\xi \geq 0.707$$

Settling time of 4 seconds from Eq. 6.3.8, gives

$$\xi\omega_n \geq 1$$

Using $\xi = 0.707$ and $\omega_n = 1/0.707$, we find

$$K = 2 \quad \text{and} \quad a = 2$$

6.4 Steady-State Response

Accuracy is measured in terms of steady state error. *Steady-state error* is the difference between the steady-state output and the desired output. Consider a unity feedback system shown in Fig. 6.9. Then the error is given by

$$e(t) = r(t) - y(t) \tag{6.4.1}$$

and the steady-state error is given by

$$e_{ss} = \lim_{t \to \infty} e(t) = \lim_{s \to 0} sE(s) \tag{6.4.2}$$

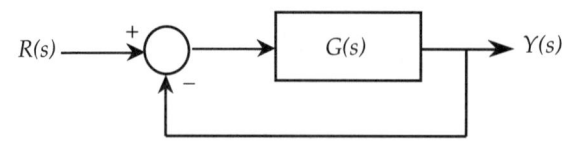

Figure 6.9 Unity feedback system.

Or using Eq.6.1.7, the error $E(s)$ can be expressed as

$$E(s) = \frac{R(s)}{1+G(s)} \qquad (6.4.3)$$

Notice that the error depends on $G(s)$ and the input $R(s)$. The error can be computed directly by using Eq.6.4.3 or by using error constants, K_p, K_v, K_a defined as

$$K_p = \lim_{s \to 0} G(s), \quad K_v = \lim_{s \to 0} sG(s), \quad K_a = \lim_{s \to 0} s^2 G(s) \qquad (6.4.4)$$

Steady state errors e_{ss} to different input signals can be obtained by using the error constants defined above:

For a step input $r(t) = Ru(t)$:

$$e_{ss} = \frac{R}{1+K_p} \qquad (6.4.5)$$

For a ramp input $r(t) = Rtu(t)$:

$$e_{ss} = \frac{R}{K_v} \qquad (6.4.6)$$

For a parabolic input $r(t) = \frac{1}{2} Rt^2 u(t)$:

$$e_{ss} = \frac{R}{K_a} \qquad (6.4.7)$$

where

$$u(t) = \begin{cases} 1 & t \geq 0 \\ 0 & t < 0 \end{cases}$$

Notice the error constants K_p, K_v, K_a depend on the number of poles at $s = 0$ for function $G(s)$.

The *type of the system* is defined as the number of poles at $s = 0$ for the transfer function $G(s)$. Consider the function

$$G(s) = \frac{(s+a)(s+b)}{s^n (s+c)^m (s^2 + ps + q)} \qquad (6.4.8)$$

The value of n determines the type of the system. For example, $n = 1$ is a type-1 system. For $n = 1$ it can be seen that $K_p = \infty$, $K_v =$ finite, $K_a = 0$. Hence the steady-state error to a step input is zero, the error due to a ramp input is finite and the error due to parabolic input is infinity.

Example 6.4

Given the following system with
$$G_1(s) = \frac{(s+6)}{s(s+2)(s+4)}$$

a) Find the value of K such that the steady-state error to a unit ramp input is 5% ($D(s) = 0$).
b) Find the steady-state error to a unit step disturbance for this value of K.

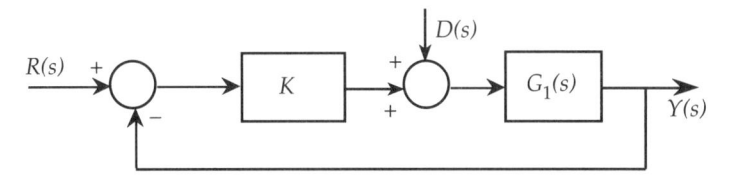

Solution: a) We see that
$$G(s) = \frac{K(s+6)}{s(s+2)(s+4)}, \quad H(s) = 1$$

This is a type-1 system:
$$K_v = 6K/8$$
$$e_{ss} = \frac{1}{K_v} = \frac{8}{6K} = 0.05. \quad \therefore K = \frac{80}{3}$$

b) It is important to note that the error due to the input is the difference between the output and the input. However, the errors due to disturbance is the actual deviation of the output from the desired inputs $R(s)$. To determine the errors due to disturbance, we can arbitrarily set $R(s) = 0$ and determine the steady-state output due to the disturbance. This represents the steady-state error due to disturbance.

Setting $R(s) = 0$, we can determine $Y(s)/D(s)$:
$$Y(s) = \frac{s+6}{s(s+2)(s+4) + K(s+6)} D(s)$$

Given $D(s) = 1/s$, we have the steady-state error due to disturbance
$$Y_{ss} = \lim_{s \to 0} sY(s) = \frac{3}{80}$$

6.5 Root Locus

Root locus method is a graphical procedure for tracing the path of the characteristic roots in the s-plane as a system parameter is changed. It was introduced by Evans in 1948.

Given the closed-loop system shown in Fig. 6.10. The characteristic equation is given by

$$1 + KG(s)H(s) = 0 \quad \text{or} \quad KG(s)H(s) = -1 \qquad (6.5.1)$$

where K is the variable parameter and varies between $-\infty$ and $+\infty$. The location of the roots in the s-plane can be varied by changing the value of K. Since s is a complex variable, Eq.6.5.1 can be written as

$$|KG(s)H(s)| = 1 \qquad \text{(Condition I)} \qquad (6.5.2)$$

and

$$\angle KG(s)H(s) = 180° + 2r\pi \qquad \text{(Condition II)} \qquad (6.5.3)$$

where $r = 0, \pm 1, \pm 2 \cdots$.

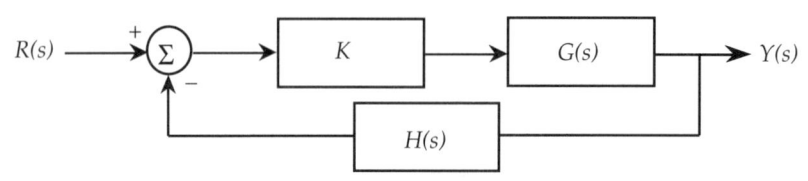

Figure 6.10 Closed-loop system.

Condition I: It is the same for all values of K (positive or negative).

Condition II: For $K > 0$ $\angle KG(s)H(s) = (2r+1)\pi$ (odd multiples of π)

For $K < 0$ $\angle KG(s)H(s) = 2r\pi$ (even multiples of π)

Several rules are derived based on the conditions given above. They are given next.

1. The points at which the root locus starts ($K = 0$) are at the poles of $G(s)H(s)$.

2. The points at which the root locus ends ($K = \pm\infty$) are at the zeros of $G(s)H(s)$.

6. The total number of root loci for Eq.6.5.1 equals the greater of the number of finite poles P and finite zeros Z of $G(s)H(s)$.

4. The root locus is symmetrical with respect to the real axis.

5. The root loci proceed to the zeros at infinity along the asymptotes centered at σ and with angles ϕ given by:

$$\phi = \frac{(2r+1)\pi}{P-Z}, \quad r = 0,1,2,\cdots,P-Z-1 \quad \text{for } K > 0$$

$$\phi = \frac{2r\pi}{P-Z}, \quad r = 0,1,2,\cdots,P-Z-1 \quad \text{for } K < 0 \qquad (6.5.4)$$

$$\sigma = \frac{\sum \text{Poles} - \sum \text{Zeros}}{P-Z} \qquad (6.5.5)$$

6. The existence of the root locus on the real axis can be determined by counting the number of poles and zeros of $G(s)H(s)$ on the real axis. When $K > 0$, the root loci are found on sections where the total number of poles and zeros to the right of any point in that section is odd. For $K < 0$, the condition is even.

7. The intersection of the root locus on the imaginary axis is found by using the Routh-Hurwitz criterion.

8. Breakaway points of the root loci represent multiple roots for the value of K. One way to determine these points is by solving the equation $dK/ds = 0$ where $K = -[G(s)H(s)]^{-1}$. It is important to note that this is only a necessary condition.

9. The angles of departure from the complex poles and the angles of arrival at the complex zeros are determined using the angle criterion. This is done by selecting a point close the complex pole or zero and measuring angles from all the finite poles and zeros of $G(s)H(s)$.

10. The magnitude of K at any point on the root locus is obtained by using the magnitude condition.

These rules are illustrated with the following example.

Example 6.5

A negative feedback control system has the following loop transfer function:

$$K G(s)H(s) = \frac{K}{s(s+1)(s^2+4s+13)}$$

Draw root locus for $K > 0$.

Solution: We will use the rules given above.

The system has no finite zeros. The poles are at $s = 0, -1, -2+j3, -2-j3$. Hence, $Z = 0$ and
$P = 4$.

There are four zeros at infinity.

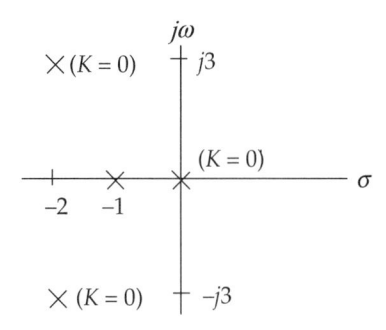

Figure E6.5a Plot of the poles and zeros.

There are 4 asymptotes. The angles and intersection point are now determined.

Angles: $\phi = \frac{(2r+1)\pi}{4}$, $r = 0, 1, 2, \cdots$ and are given by: $45°, 135°, 225°, 315°$

Intersection point: $\sigma = \dfrac{\sum \text{Poles} - \sum \text{Zeros}}{P - Z} = \dfrac{0 - 1 - 2 + j3 - 2 - j3}{4} = -1.25$

These are shown in Fig. E6.5b.

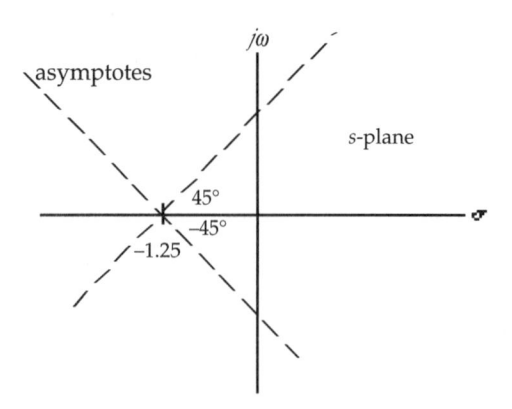

Figure E6.5b Asymptotes.

The root loci start at the poles. The root locus does not exist on the real axis between -1 and $-\infty$ or 0 and ∞ as shown in Fig. E6.5c.

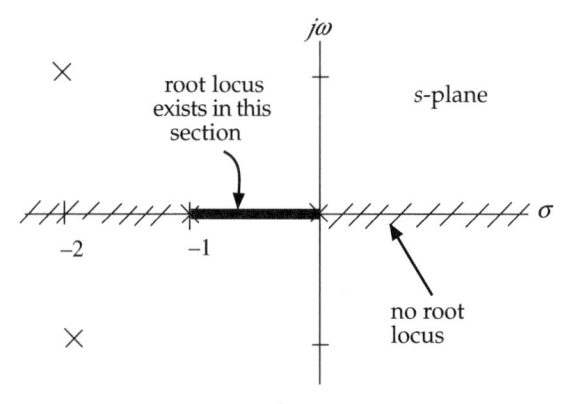

Figure E6.5c Existence of root locus on the real axis.

The <u>breakaway points</u> are obtained from $dK/ds = 0$ where K is given by
$$K = -(s^4 + 5s^3 + 17s^2 + 13s)$$
Solving this, we get
$$s = -0.467, \; s = -1.642 \pm j2.063$$

We can see from the root locus plot, only $s = -0.467$ qualifies as a breakpoint.

The angle of departure from the complex pole is obtained by using the angle condition. This is done by selecting a point close to the complex pole and measuring angles from all the finite poles and zeros of $G(s)H(s)$. This is shown in Fig. E6.5d:
$$-124° - 108° - 90° - \theta = 180° \quad \text{gives} \quad \theta = -142°$$

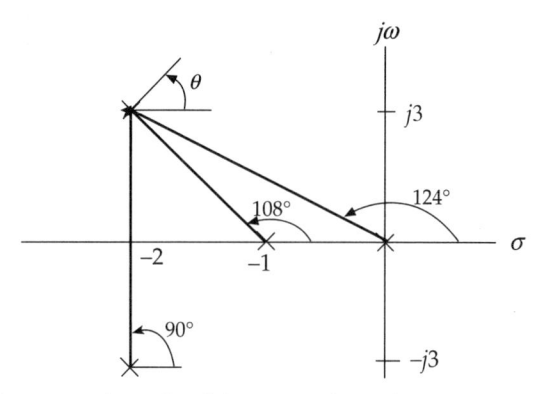

Figure E6.5d Angle of departure from the complex poles.

The Routh-Hurwitz criterion is used to find the intersection where the root loci crosses the imaginary axis. The characteristic equation is

$$s^4 + 5s^3 + 17s^2 + 13s + K = 0$$

s^4	1	17	K
s^3	5	13	
s^2	14.4	K	
s^1	$13 - 5K/14.4$	0	
s^0	K		

From the above table, the system is stable if $0 < K < 37.4$.

The intersection of the root locus on the imaginary axis is $\pm j1.6124$. The root locus plot is shown in Fig. E6.5e.

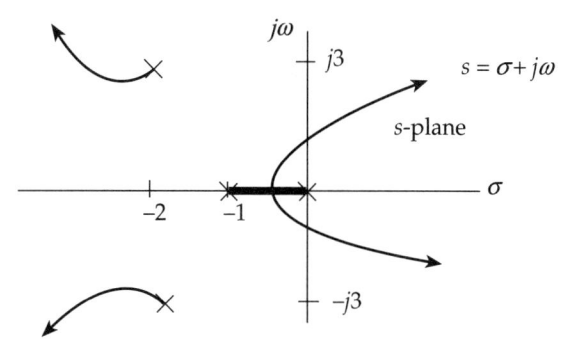

Figure E6.5e Root locus plot.

6.6 Frequency Response

Frequency response of a system is defined as the steady-state response of a system to periodic inputs. Given a transfer function $G(s)$, then the steady-state output $y(t)$ for a sinusoidal input signal $r(t) = A\sin\omega t$ is given by (see Fig. 6.16)

$$y(t) = A|G(j\omega)|\sin(\omega t + \angle G(j\omega)) \quad (6.6.1)$$

where $|G(j\omega)|$ and $\angle G(j\omega)$ represent the magnitude and angle of $G(j\omega)$, respectively. $G(j\omega)$ is called the *frequency response function*. The above result states that the magnitude and phase angle of the output as a function of frequency ω for a system with a transfer function $G(s)$ can be determined simply by replacing s by $j\omega$. The frequency response can also be obtained in the lab.

$$A\sin\omega t \longrightarrow \boxed{G(s)} \longrightarrow A|G(j\omega)|\sin(\omega t + \angle G(j\omega))$$

Figure 6.11 Output for a sinusoidal input.

A typical frequency response is shown in Fig. 6.12 where, the magnitude is typically plotted in decibels.

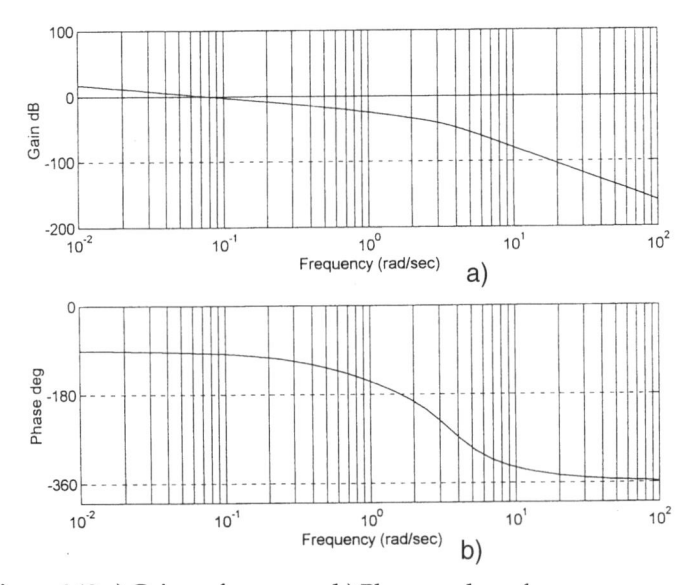

Figure 6.12 a) Gain vs frequency. b) Phase angle vs frequency.

Closed-loop stability can be determined by knowing the frequency response of the loop transfer function $G(s)H(s)$. The closed-loop system is stable if at a frequency $\omega = \omega_g$ where $|G(j\omega_g)H(j\omega_g)| = 1$ (0 db), the phase angle $\angle G(j\omega_g)H(j\omega_g)$ is less negative than $-180°$. This phase angle at this frequency also gives us the *phase margin*—the maximum negative phase shift (delay) that can be added before the system becomes unstable. The frequency ω_g is called the *gain-cross over frequency*. Closed-loop stability can also be determined by knowing the gain at the frequency $\omega = \omega_c$ where $\angle G(j\omega_c)H(j\omega_c) = -180°$. The system is stable if $|G(j\omega_c)H(j\omega_c)| < 1$ (below the 0 dB line). The gain at this frequency also provides the measure of *gain margin*—the gain that can be added to the system before it becomes unstable. The frequency ω_c is called the *phase-crossover frequency*. For phase margins of $< 70°$, the following relationship can be used to determine the approximate damping ξ which relates to percent overshoot to a step input in the time domain:

$$\xi = 0.01 \times \text{phase margin (in degrees)} \qquad (6.6.2)$$

Effect of Gain on the Frequency Response

Increasing the gain of the system will effect only the magnitude part of the frequency response of the system. This has the effect of changing both the gain and phase margins. The effect is illustrated in Fig. 6.13.

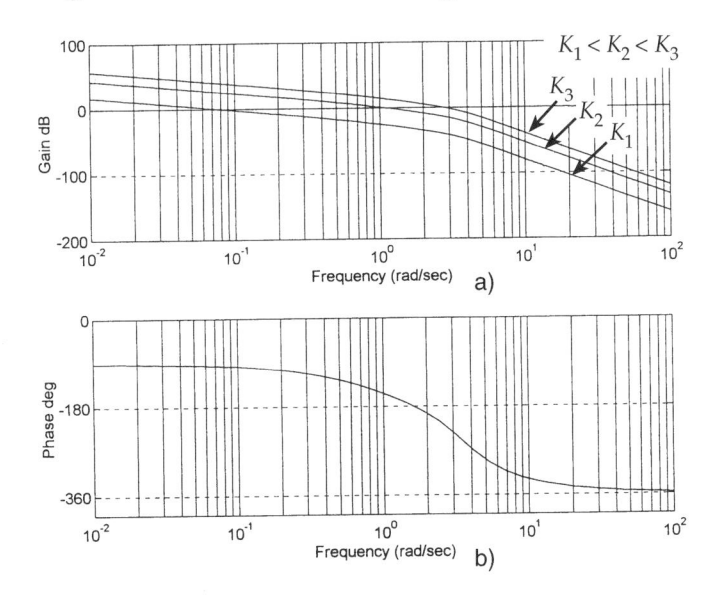

Figure 6.13 a) Gain vs frequency. b) Phase angle vs frequency.

Effect of Pure Time Delay on the Frequency Response

The transfer function of pure time delay is given by $e^{-j\omega T}$ where T is the time delay. This term has a magnitude of one and phase angle of $-\omega T$ radians. Thus, the magnitude part of the frequency response plot does not change. On the other hand, the phase plot is shifted down by an angle $\omega T \times 180/\pi$ degrees, as shown in Fig. 6.14. Notice the phase lag increases as the frequency increases for a given time delay T. This has the effect of reducing the phase margin and thus too much delay can cause a system to become unstable.

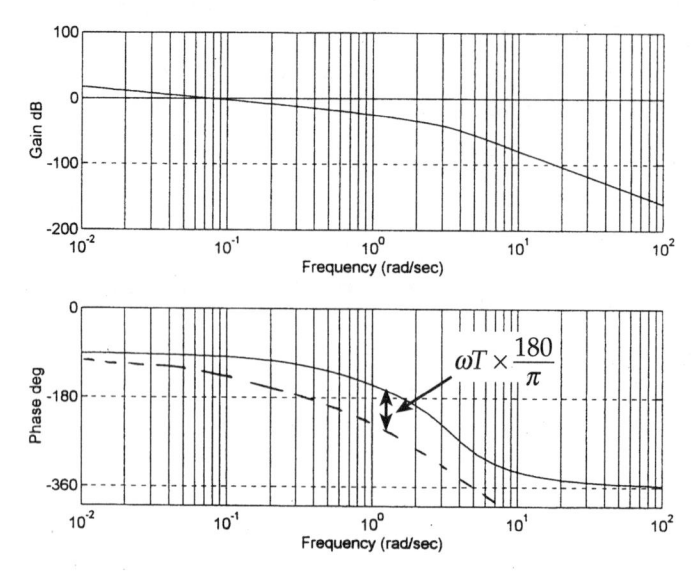

Figure 6.14 a) Gain vs frequency. b) Phase angle vs frequency.

Control Systems Practice Problems

Questions 6.1–6.9

A control system is described by the following block diagram:

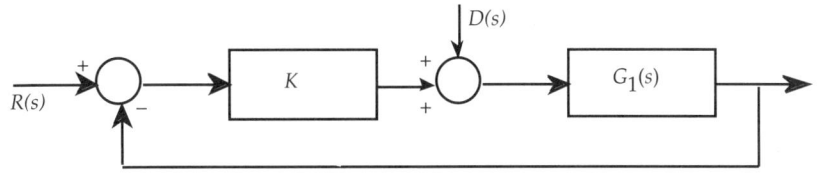

Given:
$$G_1(s) = \frac{1}{s(s+5)(s+10)}$$

Frequency response data for $G(s)$, with $K = 100$:

Frequency	magnitude	magnitude dB	phase ∠
1.389	1.374	2.757	–113.4
1.526	1.239	1.860	–115.7
1.677	1.115	0.9475	–118.1
1.842	1.002	0.0168	–120.0
1.931	0.9488	–0.4562	–122.0
2.560	0.6738	–3.429	–131.5
4.00	0.3497	–9.127	–151.6
5.00	0.250	–11.76	–160.0
5.96	0.1851	–14.65	–170.0
6.551	0.1549	–16.20	–175.9
7.197	0.1287	–17.81	–180.0
7.906	0.1061	–19.49	–186.0
8.685	0.0867	–21.00	–191.0

6.1 The value of K at $s = -6$ on the root locus for the system is:

(A) –10
(B) 20
(C) –24
(D) –32

6.2 For values of $K < 0$, the root locus has a breakaway point:

(A) between 0 and ∞
(B) between 0 and –5
(C) between –5 and –10
(D) between –10 and –∞

6.3 If one of the roots of the closed-loop system is at $s = -12$, where are the other two roots?

(A) $-5, -10$
(B) $0, -5$
(C) $-1.5 + j3.427, -1.5 - j3.427$
(D) $-2 + j5, -2 - j5$

6.4 The range of K for which the system is stable is given by:

(A) All values $K > 0$
(B) $0 < K < 750$
(C) All values of $K < 0$
(D) Unstable for all values of K

6.5 If $K = 10$, the steady state value of the output due to a disturbance $(R(s) = 0)$ is given by:

(A) 10
(B) 0.1
(C) 0.5
(D) 0.16

6.6 If $K = 10$, the sensitivity of the system to K is given by:

(A) $\dfrac{10}{s(s+5)(s+10)}$

(B) $\dfrac{10}{s(s+5)(s+10)+10}$

(C) $\dfrac{s(s+5)(s+10)}{s(s+5)(s+10)+10}$

(D) $(s+5)(s+10)+10s$

6.7 If for some value of K, the system has dominant roots at $-1.5 \pm j3.4$, the approximate relative damping and the settling time for a step input are:

(A) 0.5 and 5 sec
(B) 0.4 and 4 sec
(C) 0.4 and 2.67 sec
(D) 0.707 and 3.4 sec

6.8 The gain margin and the gain cross-over frequency are:

(A) 18 dB and 3.2 rad/s
(B) 20 dB and 1.84 rad/s
(C) 18 dB and 1.84 rad/s
(D) infinity and infinity

6.9 The phase margin and the phase cross-over frequencies are:

(A) 60° and 1.84 rad/s
(B) 50° and 5 rad/s
(C) 60° and 3 rad/s
(D) 60° and 7.19 rad/s

Questions 6.10-6.18

A control system is described by the following block diagram:

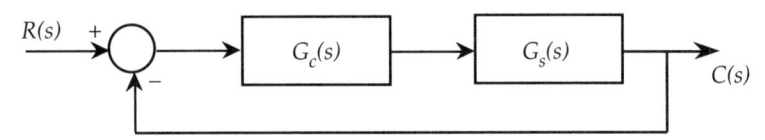

Given:

The system transfer function is

$$G_s(s) = \frac{s+4}{s(s^2+4s+13)}$$

The controller transfer function is

$$G_c(s) = \frac{K(s+a)}{s+1}$$

Possible values of interest are for $a = 3$ or 5.

6.10 The open-loop transfer function is:

(A) K

(B) $\dfrac{K(s+a)}{s+1}$

(C) $\dfrac{K(s+a)(s+4)}{s(s+1)(s^2+4s+13)}$

(D) $\dfrac{(s+4)}{s(s^2+4s+13)}$

6.11 The poles of the open-loop transfer function are:

(A) $-a, -4$ only
(B) $0, -1, -2 \pm j3$
(C) $0, -1$ only
(D) $\infty, \infty, 0, -1$

6.12 The zeros of the open-loop transfer function are:

(A) $-a, -4, \infty, \infty$
(B) $0, -1$
(C) $-2 \pm j3, \infty, \infty$
(D) $-3 \pm j2$

6.13 The number of asymptotes of the root locus for this system are:

(A) 3
(B) 2
(C) 1
(D) 0

6.14 The intersection of the asymptotes of the root locus with the real axis of the s-plane if $a = 3$ and $a = 5$ are:

(A) 1 and 2
(B) 0 and 2
(C) −2 and 1
(D) −2 and −1

6.15 The angles (in degrees) made by the asymptotes of the root locus with the real axis are:

(A) ±0
(B) ±90
(C) ±45, 180
(D) ±60, ±30

6.16 If we set $a = 5$, the angle of departure of the root locus from the complex root $s = -2 + j3$ is:

(A) 90°
(B) 180°
(C) −41°
(D) 49°

6.17 The characteristic equation of the closed-loop system for $a = 3$ is:

(A) $s^4 + 5s^3 + 13s^2 + Ks + 4K = 0$
(B) $s^4 + 5s^3 + (13+K)s^2 + Ks + 4K = 0$
(C) $s^4 + 5s^3 + (17+K)s^2 + (13+7K)s + 12K = 0$
(D) $s^4 + 5s^3 + 13s^2 + Ks + K = 0$

6.18 The gain K at $s = -0.5$ on the root locus, if $a = 3$ is:
- (A) 0.21
- (B) 0.321
- (C) −0.321
- (D) 0.98

Solutions to Practice Problems

6.1 **(C)** From the given transfer function, we can see that the existence of root is only possible for $K < 0$. Substituting $s = -6$ in the characteristic equation $s(s+5)(s+10) + K = 0$ gives $K = -24$.

6.2 **(B)** The poles are at −10, −5, and 0. The pole at $s = -10$ goes toward negative infinity. The root locus can only exist between 0 and −5. hence, the breakaway point lies between 0 and −5.

6.3 **(C)** Substituting $s = -12$ in the characteristic equation gives $K = 168$. Dividing the characteristic equation by $(s+12)$ gives a quadratic equation. The roots of this quadratic are $-1.5 \pm j3.427$.

6.4 **(B)** Use Routh-Hurwitz criterion.

6.5 **(B)** The transfer function between $C(s)$ and $D(s)$ is given by
$$\frac{C(s)}{D(s)} = \frac{1}{s(s+5)(s+10) + K}$$

Applying the final value theorem $\lim_{s \to 0} sC(s)$ and substituting $K = 10$, we get $C_{ss} = 0.1$.

6.6 **(C)** Apply the sensitivity formula Eq. 6.2.6 with $G(s) = \dfrac{K}{s(s+5)(s+10)}$ and $H(s) = 1$:
$$S_G^T = \frac{1}{1+GH} = \frac{s(s+5)(s+10)}{s(s+5)(s+10) + 10}$$

6.7 **(C)** Use Eqs. 6.3.4 and 6.3.8 and refer to Fig. 6.8:
$$\xi \omega_n = 1.5, \quad \omega_n\sqrt{1-\xi^2} = 3.4. \quad \therefore \xi = 0.4, \; t_s = \frac{4}{1.5} = 2.67$$

6.8 **(C)** From the table, the approximate gain margin and the gain cross-over frequency are 18 dB and 1.84 rad/s. respectively.

6.9 **(D)** The approximate phase margin and the phase cross-over frequency are 60 degrees and 7.19 rad/s.

6.10 **(C)** $G_c(s)G_s(s) = \dfrac{K(s+a)}{s+1} \dfrac{s+4}{s(s^2+4s+13)}$

6.11 **(B)** Poles are given by the values of s that make $G_c(s)G_s(s) = \infty$.

6.12 **(A)** Zeros are given by the values of s that make $G_c(s)G_s(s) = 0$.

6.13 **(B)** There are two zeros at $s = \infty$ since as $s \to \infty$, $\dfrac{Ks^2}{s^4} = \dfrac{K}{s^2}$. Hence, there are two asymptotes.

6.14 **(A)** With $a = 3$, $\sigma = \dfrac{0-1-2+j3-2-j3-(-3-4)}{4-2} = 1$. With $a = 5$, $\sigma = 2$.

6.15 **(B)** With $P - Z = 4 - 2 = 2$, $r = 0$ and 1. Then

$\phi = \dfrac{(2\times 0+1)\pi}{2} = \dfrac{\pi}{2}$ and $\phi = \dfrac{(2\times 1+1)\pi}{2} = \dfrac{3\pi}{2}$ or $-\dfrac{\pi}{2}$

6.16 **(C)** $180° = -\theta - \angle s + \angle(s+4) + \angle(s+5) - \angle(s+1) - \angle(s+2+j3)$
$= -\theta - 123.7° + 56.3° + 45° - 109° - 90°$. $\therefore \theta = -41°$

6.17 **(C)** The characteristic equation is $1 + G_c(s)G_s(s) = 0$. Substitute $a = 6$.

6.18 **(B)** Use Eq. 6.5.2 with $s = -0.5$:

$\left| \dfrac{-0.5+4}{-0.5[(-0.5)^2+4(-0.5)+13]} \times \dfrac{K(-0.5+3)}{-0.5+1} \right| = 1$ $\therefore K = 0.321$

Electrical Engineering Discipline Exam

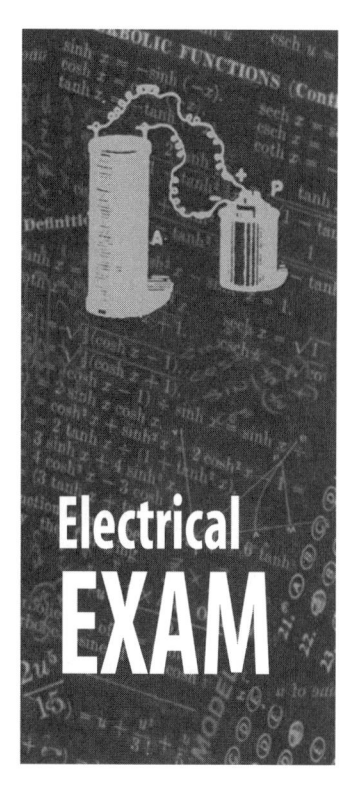

This practice exam has been developed to help test-takers prepare for the afternoon Discipline-Specific test in Electrical Engineering. In addition, if test-takers are undecided, taking this test may also help them determine whether they should take the afternoon DS test or the afternoon General test.

The subjects tested are those subjects that the NCEES says will make up the exam. We have placed the same number of problems from each area of Electrical Engineering that can be expected to be on the actual exam. The difficulty level can, of course, be quite different on the actual exam; we have, however, attempted to create an exam with essentially the same level of difficulty as one would expect on the actual exam.

If you find the problems on this exam to be more difficult than the problems on the afternoon General test, you may wish to take the General test in the afternoon. About one-half of all Electrical Engineering examinees take the afternoon General test. This practice exam should help you make that decision before the day of the test.

This exam is also included on the *free* CD attached to the back cover of this book. The Study-Director™ feature of the CD will provide an analysis of your results and help you determine which subjects to review again, should you need extra review. Do not look at this exam if you plan to take the exam on the CD.

FUNDAMENTALS OF ENGINEERING EXAM

Afternoon Session—Electrical Exam

(Simulated answer form with topical breakout and scoring grid.)

BE SURE EACH MARK IS DARK AND COMPLETELY FILLS THE INTENDED SPACE AS ILLUSTRATED HERE: ●.

POWER SYSTEMS
1 Ⓐ Ⓑ Ⓒ Ⓓ
2 Ⓐ Ⓑ Ⓒ Ⓓ
3 Ⓐ Ⓑ Ⓒ Ⓓ

Score: _____

SIGNAL PROCESSING
16 Ⓐ Ⓑ Ⓒ Ⓓ
17 Ⓐ Ⓑ Ⓒ Ⓓ
18 Ⓐ Ⓑ Ⓒ Ⓓ

Score: _____

ANALOG CIRCUITS
25 Ⓐ Ⓑ Ⓒ Ⓓ
26 Ⓐ Ⓑ Ⓒ Ⓓ
27 Ⓐ Ⓑ Ⓒ Ⓓ
28 Ⓐ Ⓑ Ⓒ Ⓓ
29 Ⓐ Ⓑ Ⓒ Ⓓ
30 Ⓐ Ⓑ Ⓒ Ⓓ

Score: _____

ELECTROMAGNETICS
34 Ⓐ Ⓑ Ⓒ Ⓓ
35 Ⓐ Ⓑ Ⓒ Ⓓ
36 Ⓐ Ⓑ Ⓒ Ⓓ
37 Ⓐ Ⓑ Ⓒ Ⓓ
38 Ⓐ Ⓑ Ⓒ Ⓓ
39 Ⓐ Ⓑ Ⓒ Ⓓ

Score: _____

NETWORKS
46 Ⓐ Ⓑ Ⓒ Ⓓ
47 Ⓐ Ⓑ Ⓒ Ⓓ
48 Ⓐ Ⓑ Ⓒ Ⓓ
49 Ⓐ Ⓑ Ⓒ Ⓓ
50 Ⓐ Ⓑ Ⓒ Ⓓ
51 Ⓐ Ⓑ Ⓒ Ⓓ

Score: _____

DIGITAL SYSTEMS
4 Ⓐ Ⓑ Ⓒ Ⓓ
5 Ⓐ Ⓑ Ⓒ Ⓓ
6 Ⓐ Ⓑ Ⓒ Ⓓ
7 Ⓐ Ⓑ Ⓒ Ⓓ
8 Ⓐ Ⓑ Ⓒ Ⓓ
9 Ⓐ Ⓑ Ⓒ Ⓓ

Score: _____

SOLID STATE
19 Ⓐ Ⓑ Ⓒ Ⓓ
20 Ⓐ Ⓑ Ⓒ Ⓓ
21 Ⓐ Ⓑ Ⓒ Ⓓ
22 Ⓐ Ⓑ Ⓒ Ⓓ
23 Ⓐ Ⓑ Ⓒ Ⓓ
24 Ⓐ Ⓑ Ⓒ Ⓓ

Score: _____

INSTRUMENTATION
31 Ⓐ Ⓑ Ⓒ Ⓓ
32 Ⓐ Ⓑ Ⓒ Ⓓ
33 Ⓐ Ⓑ Ⓒ Ⓓ

Score: _____

COMMUNICATIONS
40 Ⓐ Ⓑ Ⓒ Ⓓ
41 Ⓐ Ⓑ Ⓒ Ⓓ
42 Ⓐ Ⓑ Ⓒ Ⓓ
43 Ⓐ Ⓑ Ⓒ Ⓓ
44 Ⓐ Ⓑ Ⓒ Ⓓ
45 Ⓐ Ⓑ Ⓒ Ⓓ

Score: _____

COMPUTERS
52 Ⓐ Ⓑ Ⓒ Ⓓ
53 Ⓐ Ⓑ Ⓒ Ⓓ
54 Ⓐ Ⓑ Ⓒ Ⓓ
55 Ⓐ Ⓑ Ⓒ Ⓓ
56 Ⓐ Ⓑ Ⓒ Ⓓ
57 Ⓐ Ⓑ Ⓒ Ⓓ
58 Ⓐ Ⓑ Ⓒ Ⓓ
59 Ⓐ Ⓑ Ⓒ Ⓓ
60 Ⓐ Ⓑ Ⓒ Ⓓ

Score: _____

CONTROLS
10 Ⓐ Ⓑ Ⓒ Ⓓ
11 Ⓐ Ⓑ Ⓒ Ⓓ
12 Ⓐ Ⓑ Ⓒ Ⓓ
13 Ⓐ Ⓑ Ⓒ Ⓓ
14 Ⓐ Ⓑ Ⓒ Ⓓ
15 Ⓐ Ⓑ Ⓒ Ⓓ

Score: _____

cut here!

Electrical Discipline Exam

Three single-phase, 7200 volt to 480 volt, 100 kVA transformers are connected in a step-down bank with the 7200 windings connected in wye and the 480 volt windings connected in delta.

Questions 1 – 3

1. What is the rated line-to-line voltage on the wye side?
    (A) 4,160 V
    (B) 7,200 V
    (C) 12,470 V
    (D) 14,400 V

2. What is the rated line current on the wye side?
    (A) 4.63 A
    (B) 13.9 A
    (C) 24.0 A
    (D) 41.6 A

3. What is the rated line current on the delta side?
    (A) 208 A
    (B) 361 A
    (C) 480 A
    (D) 625 A

A logic circuit is shown:

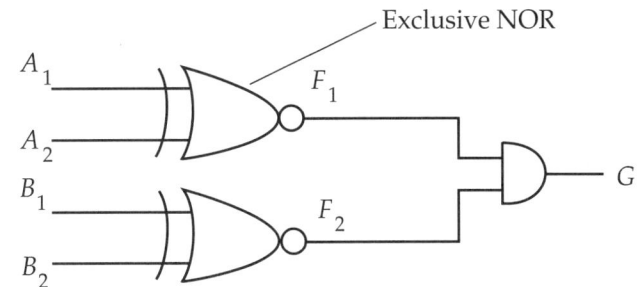

Questions 4 – 6

4. The function F_1 in the logic circuit is given by:
    (A) $A_1 A_2$
    (B) $A_1 A_2 + \overline{A_1}\overline{A_2}$
    (C) $A_1 + A_2$
    (D) $A_1 + A_1 A_2$

5. If $A_1 = 1$ and $B_1 = 1$, then $G = 1$ if:
    (A) $A_2 = 1$ and $B_2 = 1$
    (B) $A_2 = 0$ and $B_2 = 1$
    (C) $A_2 = 0$ and $B_2 = 0$
    (D) $A_2 = 1$ and $B_2 = 0$

6. If $B_1 = 1$ and $B_2 = 1$, then F_2 will be equal to
   (A) 0
   (B) 1
   (C) not defined
   (D) F_1

Questions 7 – 9

Flip-flops A and B form a sequential circuit as shown:

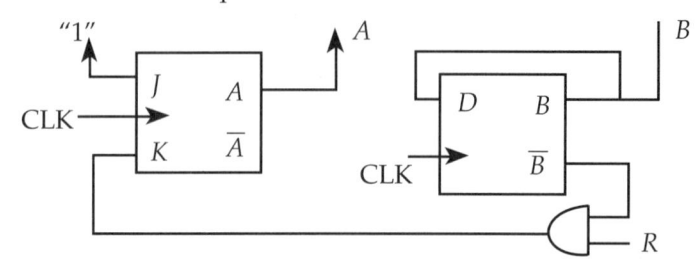

7. If $R = 1$, after the clock pulse, the binary count 10 changes to:
   (A) 00
   (B) 01
   (C) 10
   (D) 11

8. If $R = 0$, after the clock pulse the binary count 00 changes to:
   (A) 00
   (B) 01
   (C) 10
   (D) 11

9. If $R = 1$, after the clock pulse the binary count 01 changes to:
   (A) 00
   (B) 01
   (C) 10
   (D) 11

Questions 10 – 12

A servo system for a pen plotter is given by the following block diagram:

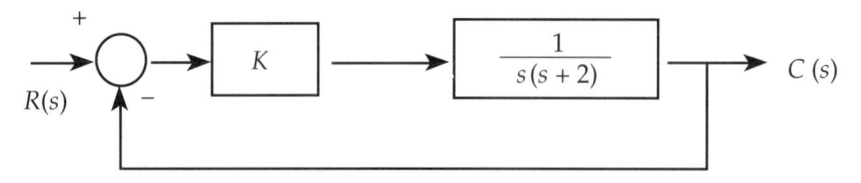

10. The value of K required to result in a closed-loop system characteristic equation of $P(s) = s^2 + 2s + 100$ is given by:
    (A) 0
    (B) 50
    (C) 100
    (D) 25

11. The value of K required to get the fastest response without any overshoot ($\zeta = 1$) is given by:
    (A) 1
    (B) 10
    (C) 5
    (D) 100

12. If K is set to 16, the damped natural frequency of oscillation to a step input is most nearly:
    (A) 4 rad/s
    (B) 16 rad/s
    (C) 12 rad/s
    (D) 10 rad/s

Given the following closed-loop system:

Questions 13 – 15

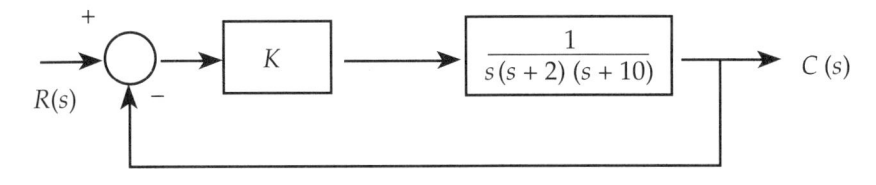

13. The value of K for which the system becomes marginally stable is nearly:
    (A) 10
    (B) 240
    (C) 100
    (D) 60

14. The frequency of oscillation when K is set to make the system marginally stable is :
    (A) 0.7 Hz
    (B) 10 Hz
    (C) 1.2 Hz
    (D) 2 Hz

15. If K = 0, the steady-state error to a step input of magnitude 10 is:
    (A) 0
    (B) 10
    (C) 2
    (D) 200

Questions 16 – 18

Two discrete-time signal processing systems are depicted below. Each system consists of a decimator and a linear time-invariant filter in series. The input to both systems is the discrete-time signal $x[n]$. The discrete-time impulse response of the linear time-invariant filter is denoted by $h[n]$. The decimator discards every odd sample from its input to produce its output; for example, the input $x[n]$ to the decimator in System I produces the output $x[2n]$.

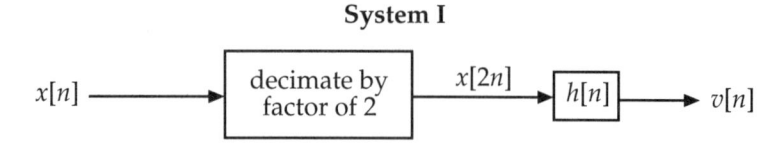

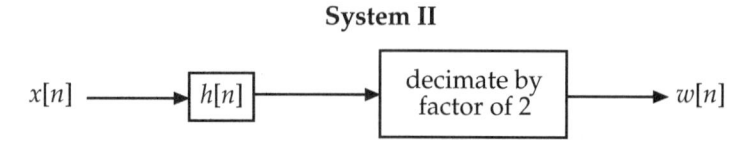

System I

$x[n] \longrightarrow$ [decimate by factor of 2] $\xrightarrow{x[2n]}$ [$h[n]$] $\longrightarrow v[n]$

System II

$x[n] \longrightarrow$ [$h[n]$] $\longrightarrow$ [decimate by factor of 2] $\longrightarrow w[n]$

16. In general, with no special assumptions on $x[n]$ or $h[n]$,
    (A) $v[n] = w[n]$, and both systems are time-invariant.
    (B) $v[n] \neq w[n]$, and both systems are time-invariant.
    (C) $v[n] = w[n]$, and both systems are time-varying.
    (D) $v[n] \neq w[n]$, and both systems are time-varying.

17. If $h[n] = \delta[n] + 0.5\delta[n-1]$, where $\delta[n]$ is a unit-sample (discrete-time impulse) function, and $x[n] = \cos \pi n$, then the output $v[n]$ of System I is:
    (A) $v[n] = 0$, a constant signal
    (B) $v[n] = 1.5$, a constant signal
    (C) $v[n] = -0.5 \cos \pi n$
    (D) $v[n] = 0.5^n u[n]$, where $u[n]$ is the unit-step function

18. If $h[n] = \delta[n] + \delta[n-1]$, where $\delta[n]$ is a unit-sample (discrete-time impulse) function, and $x[n] = \cos \pi n$, then the output $w[n]$ of System II is:
    (A) $w[n] = -0.5 \cos \pi n$, a constant signal
    (B) $w[n] = 1.5$, a constant signal
    (C) $w[n] = -0.5 \cos \pi n$
    (D) $w[n] = 0.5^n u[n]$, where $u[n]$ is the unit-step function

19. For a silicon pn-diode at room temperature, what should the applied diode voltage V_D be in order to cause the diode current i_D to reach 90% of the diode saturation current I_s?
    (A) 0.7 V
    (B) 0.5 V
    (C) 0.16 V
    (D) 0.016 V

20. Determine the base current I_B of the following BJT given $V_{BE} = 0.7$ V and $\beta = 100$.

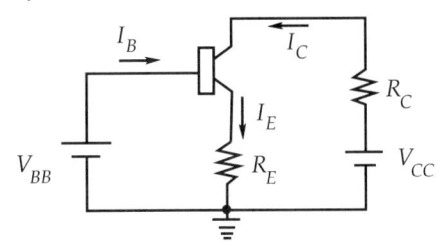

$V_{CC} = 10$ V
$V_{BB} = 5$ V
$R_C = 2$ kΩ
$R_E = 1$ kΩ

(A) 0.0426 mA
(B) 0.426 mA
(C) 0.746 mA
(D) 7.46 mA

21. The following JFET is operating in the saturation region. Find the gate-to-drain voltage V_{GD}.

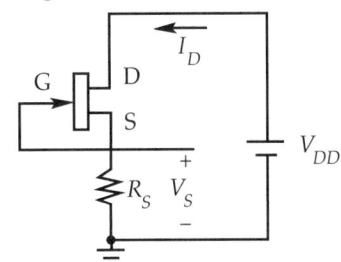

$V_{DD} = 5$ V
$R_S = 1$ kΩ
$I_{DSS} = 2$ mA
$V_p = -2$ V

(A) –2 V
(B) 2 V
(C) 0 V
(D) –3 V

22. Find the power dissipated in the JFET.

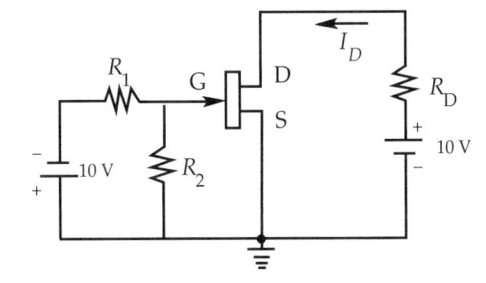

$R_1 = 90$ kΩ
$R_2 = 10$ kΩ
$R_D = 1$ kΩ
$I_{DSS} = 18$ mA
$V_p = -3$ V

(A) 0.016 W
(B) 0.16 W
(C) 1.6 W
(D) 16 W

Questions 23 – 24

Given the following BJT amplifier circuit:

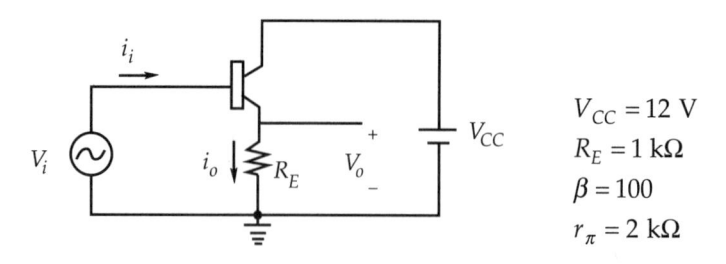

$V_{CC} = 12$ V
$R_E = 1$ kΩ
$\beta = 100$
$r_\pi = 2$ kΩ

23. Find the small-signal current gain $A_i = i_o / i_i$.
    - (A) 101
    - (B) 100
    - (C) 99
    - (D) 50

24. Find the small-signal voltage gain $A_V = V_o / V_i$.
    - (A) 0.5
    - (B) 0.98
    - (C) 9.8
    - (D) 18.8

Questions 25 – 28

For the cascade Op Amp configuration shown, respond to the following questions.

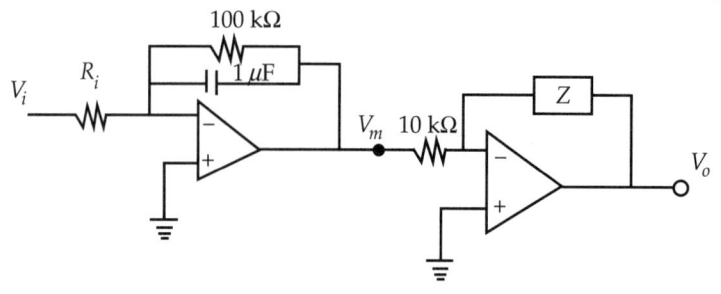

25. Determine the R_i needed to obtain an active filter with maximum amplitude gain of $V_m / V_i = 25$.
    - (A) 25 kΩ
    - (B) 10 Ω
    - (C) 2.5 kΩ
    - (D) none of the above

26. The filter bandwidth for V_m / V_i:
    - (A) cannot be determined without R_i
    - (B) is less than 2 Hz
    - (C) is greater than 10 Hz but less than 100 Hz
    - (D) is greater than 100 Hz

27. If a second-order low pass filter (2 pole filter) is desired, we can choose Z as:
    (A) a series R–C impedance with time constant, $\tau = 10^{-1}$
    (B) a parallel R–C impedance with time constant, $\tau = 10^{-1}$
    (C) a parallel R–L impedance with time constant, $\tau = 10^{-1}$
    (D) a series R–L impedance with time constant, $\tau = 10^{-1}$

28. If Z is a 1.0 henry inductor in parallel with a 10.0 Ω resistor and R_i is chosen as 1.0 kΩ, the configuration resembles:
    (A) a low-pass filter with maximum amplitude gain of 100
    (B) a high-pass filter with maximum amplitude gain of 1.0
    (C) a band-pass filter with maximum amplitude gain of 0.05
    (D) a high-pass filter with maximum amplitude gain of 100

Given the configuration shown:

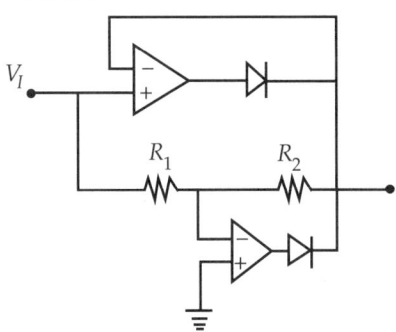

Questions 29 – 30

29. The configuration shown most nearly represents:
    (A) a full-wave rectifier
    (B) a half-wave rectifier
    (C) a voltage follower
    (D) a differential amplifier

30. If the resistor R_1 were to burn out (open circuit), the configuration would appear as:
    (A) a full-wave rectifier
    (B) a half-wave rectifier
    (C) a voltage follower
    (D) a non-inverting amplifier

Questions 31 - 33 An *npn* BJT transistor configuration is used as an instrumentation amplifier. The bias conditions are as shown. Assume $V_{BE} = 0.7$ V. This amplifier is to be used to linearly amplify signals in the range $V_i = 2.7 \pm 1.0$ V with $R_b = 2$ kΩ and $R_C = 100$ Ω.

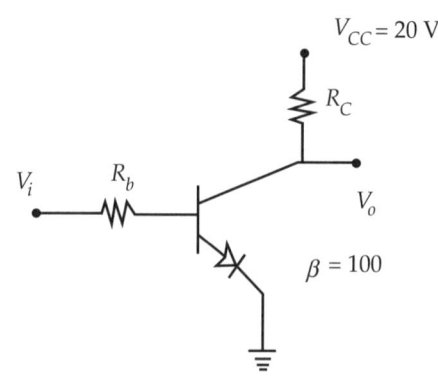

31. The value of V_o for $V_i = 1.7$ V is:
    (A) 0.5 V
    (B) 5.0 V
    (C) 10.0 V
    (D) 15.0 V

32. The value of V_o for $V_i = 3.7$ V is:
    (A) 0.5 V
    (B) 5.0 V
    (C) 10.0 V
    (D) 15.0 V

33. The value of i_e for $V_i = 2.7$ V is:
    (A) 99 mA
    (B) 100 mA
    (C) 101 mA
    (D) 102 mA

Questions 34 – 36 The field for a dipole is given by

$$E = \frac{2p\cos\theta}{4\pi\varepsilon_o r^3}\mathbf{a}_r + \frac{p\sin\theta}{4\pi\varepsilon_o r^3}\mathbf{a}_\theta$$

34. A Gaussian surface of radius r_o is taken at the origin in order to evaluate the integral $\int_A \mathbf{E} \cdot d\mathbf{A}$ for the dipole described above. The result of this surface integral is:
    (A) too complex to evaluate
    (B) $|p|$ where $|p| = Qd$, the limit as $d \to 0$
    (C) zero
    (D) $Qd/2$

35. The potential on the surface of the sphere is:
    (A) zero
    (B) constant $\neq 0$
    (C) indeterminant
    (D) $V_0 \cos\theta$ where V_0 is a constant

36. Evaluation of $\nabla \times \mathbf{E}$ and $\nabla \cdot \mathbf{E}$ would yield:
    (A) zero everywhere
    (B) zero except at $r = 0$
    (C) zero for the curl, and $|p| = Qd/\varepsilon_0$ for the divergence
    (D) zero for the divergence, and $|p| = Qd/\varepsilon_0$ for the curl

Two uniform spherical surface charge layers of equal radii and equal but opposite surface charge densities of value $\pm\rho_0$, respectively, are touching in free space, as shown. Assume the potential is referenced to zero at infinity.

Questions 37 – 39

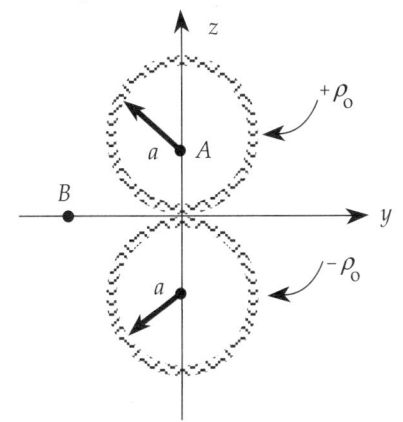

37. Find the exact potential at the point $A(0, 0, a)$.
    (A) $V = 0$
    (B) $V = -\dfrac{\rho_0}{2\varepsilon_0}$
    (C) $V = \dfrac{\rho_0}{\varepsilon_0}$
    (D) $V = \dfrac{\rho_0}{2\varepsilon_0}$

38. Find the exact potential at the point $B(0, -a, 0)$.
    (A) $V = 0$
    (B) $V = -\dfrac{\rho_0 a}{2\varepsilon_0}$
    (C) $V = \dfrac{\rho_0 a}{\varepsilon_0}$
    (D) $V = \dfrac{\rho_0 a}{2\varepsilon_0}$

39. Find the magnitude of the electric field at $B(0, -a, 0)$.
    (A) $E = 0$
    (B) $E = \dfrac{\sqrt{2}\rho_0}{2\varepsilon_0}$
    (C) $E = \dfrac{\rho_0}{2\varepsilon_0}$
    (D) $E = \dfrac{\rho_0}{\varepsilon_0}$

Questions 40 – 42

Given an AM system with modulation index $\mu = 1$ (or 100% modulation) and modulation voltage
$$v(t) = K[2\cos\omega_m t + 4\cos 2\omega_m t + 2\cos 3\omega_m t]$$
Assume $\omega_c = 2\pi f_c = 10\omega_m$.

40. The value of K for 100% modulation (without distortion) is:
    (A) 1.0
    (B) 0.5
    (C) 0.25
    (D) 0.125

41. The highest frequency carried by this AM signal is:
    (A) 10 f_m
    (B) 12 f_m
    (C) 13 f_m
    (D) 14 f_m

42. The approximate average power carried by the sidebands is:
    (A) 0.5 W
    (B) 0.1 W
    (C) 0.6 W
    (D) 1.0 W

Questions 43 – 45

Given the tone modulating signal
$$x(t) = \cos 6\pi \times 10^3 t$$
and the carrier
$$v(t) = A_c \cos 600\pi \times 10^3 t$$
with $\mu = 1$ (the modulation index).

43. The maximum frequency in this AM signal is:
    (A) 300 kHz
    (B) 600 kHz
    (C) 2 kHz
    (D) none of these

44. The power dissipated across a resistor R due to the modulated signal is:

    (A) $3A_c^2/4R$

    (B) $A_c^2/2R$

    (C) A_c^2/R

    (D) none of these

45. The signal resulting from USSB-SC (upper single sideband suppressed carrier) is:

    (A) $0.5A_c \cos 606\pi \times 10^3 t$

    (B) $A_c \cos 600\pi \times 10^3 t$

    (C) $A_c \cos 6\pi \times 10^3 t$

    (D) none of these

A switch is closed at $t = 0$ in the circuit shown.

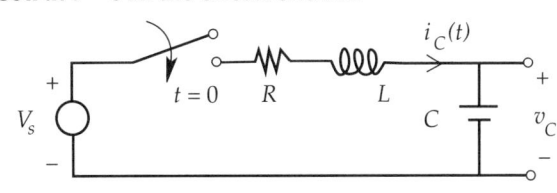

Questions 46 – 49

46. Assuming that the capacitor was initially uncharged, the current that flows at $t = 0$ after the switch is closed is:

    (A) V_s/R

    (B) V_s/L

    (C) zero

    (D) none of the above

47. The rate of change of current at $t = 0$ after the switch is closed, i.e., $i'_C(0^+)$ is:

    (A) V_s/R

    (B) V_s/L

    (C) zero

    (D) none of the above

48. The condition for an overdamped response is:

    (A) $\left(\dfrac{R}{2L}\right)^2 = \dfrac{1}{LC}$

    (B) $\left(\dfrac{R}{2L}\right)^2 > \dfrac{1}{LC}$

    (C) $\left(\dfrac{R}{2L}\right)^2 < \dfrac{1}{\sqrt{LC}}$

    (D) none of the above

49. The solution $i_c(t) = Ke^{-at}\cos(bt + \phi)$ where K, a, b, ϕ are constants:
    (A) cannot possibly exist
    (B) exists only for overdamped conditions
    (C) would require that $\phi = \pi/2$
    (D) is the solution for the given circuit

Questions 50 – 51 A certain system has two measurements performed on a set of terminals. The first provides a voltage of 20 V across a 2 kΩ resistor placed across the terminals. The second provides 25 V across a 5 kΩ resistor placed across the terminals.

50. This system has the characteristic:
    (A) no Thevenin equivalent is possible
    (B) the system is nonlinear
    (C) more information is required to make a decision
    (D) a Norton equivalent is possible

51. A 4 kΩ resistor is placed across the terminals. Assuming the system is linear, this would give:
    (A) a power dissipation of 144 mW in the resistor
    (B) a Thevenin equivalent for the system of 30 V in series with 1 kΩ
    (C) a Norton equivalent is possible
    (D) all of the above

Questions 52 – 54 A simple byte oriented microcontroller has a 16-bit stack pointer (SP) register that points to the next empty memory location in the stack. A PUSH instruction loads an operand into the stack and then decrements the SP; a PULL instruction increments the SP and then unloads an operand from the stack.

52. Based on the above description, which of the following statements about the stack is true?
    (A) The stack will fill up toward either lower memory or higher memory addresses depending on the number of items stored.
    (B) The stack will always fill up toward higher memory addresses.
    (C) The stack will always fill up toward lower memory addresses.
    (D) This stack structure will not work since the stack pointer is sometimes incremented and sometimes decremented.

53. If the SP is set initially at memory location 0000_{16}, what will it contain after the serial execution of ten successive PUSH instructions?
    (A) 0010_{16}
    (B) $000A_{16}$
    (C) $FF10_{16}$
    (D) $FFF6_{16}$

54. Which of the following is NOT a common use for a memory stack of the type described above?
    (A) The stack can be used to store program counter return links for nested subprograms.
    (B) The stack can be used to store parameters passed among subprograms.
    (C) The stack can be used to store and retrieve randomly accessed database records.
    (D) The stack can be used to sequentially store synchronously received input data.

Questions 55 – 57

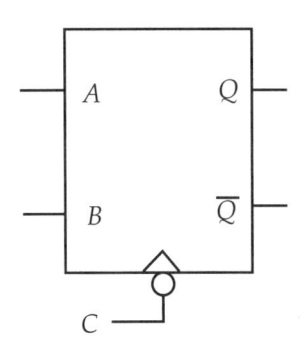

A special flip-flop, specified below, has been designed for a clocked sequential logic circuit.

A	B	Q_{n-1}
0	0	$\overline{Q}_n$
0	1	1
1	0	0
1	1	Q_n

Circuit Icon   Truth Table

55. Which of the following is a true statement about when the flip-flop output state can change?
    (A) State changes are 1-level sensitive.
    (B) State changes are 0-level sensitive.
    (C) State changes are rising-edge sensitive.
    (D) State changes are falling-edge sensitive.

56. The specified flip-flop has the following characteristic equation:
    (A) $Q_{n+1} = \overline{A}\,\overline{Q}_n + BQ_n$
    (B) $Q_{n+1} = A\overline{Q}_n + \overline{B}Q_n$
    (C) $Q_{n+1} = AB + \overline{Q}_n$
    (D) Cannot be determined from the information given.

57. Determine the excitation table for the flip-flop.

(A)

$Q_n \to Q_{n+1}$		A	B
0	0	X	1
0	1	X	0
1	0	0	X
1	1	1	X

(B)

$Q_n \to Q_{n+1}$		A	B
0	0	1	X
0	1	0	X
1	0	X	0
1	1	X	1

(C)

$Q_n \to Q_{n+1}$		A	B
0	0	1	1
0	1	0	0
1	0	1	0
1	1	0	1

(D) Cannot be determined from the information given.

58. A spreadsheet has cell $B2$ set to 4. Then cell $B3$ is set to the formula $\$B\$2 + B2$. This formula is copied into cells $B4$, $B5$, $B6$, and $B7$ in order. The number in $B6$ is:

(A) 20
(B) 24
(C) 32
(D) 16

59. The pseudo-code segment is given the following list of integer values as data: 17, 15, 12, 18, 19, 25. After execution, what values will be displayed?

```
a := 100; b := 0; c := 0;
for i := 1 to 5
  read( x );
  if( x < a )
    a := x;
  if( x > b )
    b := x;
  c := c + x;
endfor
print( a, b, c/5 );
```

(A) 12, 25, 21
(B) 12, 17, 16
(C) 12, 19, 16
(D) 12, 19, 21

60. A file contains the integers 3, 2, –4, 5, 1, –2, ... which are processed by the pseudo-code program:

```
a := 0;
i := 0;
while( i ≤ 3)
   read(b);
   if( b > 0 )
      a = a + b*b;
   i := i + 1;
endwhile;
print( a/i);
```

What will be the approximate value printed?
(A) 11
(B) 11.25
(C) 7.8
(D) 9.5

FUNDAMENTALS OF ENGINEERING EXAM

Afternoon Session—Electrical Exam

(Simulated answer form with topical breakout and scoring grid.)

BE SURE EACH MARK IS DARK AND COMPLETELY FILLS THE INTENDED SPACE AS ILLUSTRATED HERE: ●.

POWER SYSTEMS

1. (A) (B) ● (D)
2. (A) ● (C) (D)
3. (A) ● (C) (D)

Score: _____

SIGNAL PROCESSING

16. (A) (B) (C) ●
17. (A) ● (C) (D)
18. ● (B) (C) (D)

Score: _____

ANALOG CIRCUITS

25. (A) (B) (C) ●
26. (A) ● (C) (D)
27. (A) ● (C) (D)
28. (A) (B) ● (D)
29. ● (B) (C) (D)
30. (A) ● (C) (D)

Score: _____

ELECTROMAGNETICS

34. (A) (B) ● (D)
35. (A) (B) (C) ●
36. (A) ● (C) (D)
37. (A) (B) (C) ●
38. ● (B) (C) (D)
39. (A) ● (C) (D)

Score: _____

NETWORKS

46. (A) (B) ● (D)
47. (A) ● (C) (D)
48. (A) ● (C) (D)
49. (A) (B) ● (D)
50. (A) (B) ● (D)
51. (A) (B) (C) ●

Score: _____

DIGITAL SYSTEMS

4. (A) ● (C) (D)
5. ● (B) (C) (D)
6. (A) ● (C) (D)
7. ● (B) (C) (D)
8. (A) (B) ● (D)
9. (A) (B) (C) ●

Score: _____

SOLID STATE

19. (A) (B) (C) ●
20. ● (B) (C) (D)
21. (A) (B) (C) ●
22. ● (B) (C) (D)
23. ● (B) (C) (D)
24. (A) ● (C) (D)

Score: _____

INSTRUMENTATION

31. (A) (B) (C) ●
32. (A) ● (C) (D)
33. (A) (B) ● (D)

Score: _____

COMMUNICATIONS

40. (A) (B) (C) ●
41. (A) (B) ● (D)
42. (A) ● (C) (D)
43. (A) (B) (C) ●
44. ● (B) (C) (D)
45. ● (B) (C) (D)

Score: _____

COMPUTERS

52. (A) (B) ● (D)
53. (A) (B) (C) ●
54. (A) (B) ● (D)
55. (A) (B) (C) ●
56. ● (B) (C) (D)
57. (A) ● (C) (D)
58. ● (B) (C) (D)
59. (A) (B) ● (D)
60. (A) (B) (C) ●

Score: _____

CONTROLS

10. (A) (B) ● (D)
11. ● (B) (C) (D)
12. ● (B) (C) (D)
13. (A) ● (C) (D)
14. ● (B) (C) (D)
15. ● (B) (C) (D)

Score: _____

Solutions to Electrical Test

1. **C** For a wye connection,
$$V_{line} = \sqrt{3} \times V_{phase}$$
which is
$$V_{line} = 1.732 \times 7,200 = 12,470 \text{ V}$$

2. **B** We know in a three phase circuit,
$$\text{Total rated } VA = \sqrt{3} \times V_{line} \times I_{line}$$
from which
$$I_{line} = \frac{VA}{\sqrt{3} \times V_{line}}$$
and also that the Total rated $VA = 3 \times$ rated VA of each transformer:
$$\text{Total rated } VA = 3 \times 100,000 = 300,000$$
from which
$$I_{line} = 300,000 / (\sqrt{3} \times 12,470) = 13.9 \text{ A}$$

3. **B** This is identical to Problem 2, but we must realize that on the delta side,
$$V_{line} = V_{phase}, \quad \text{or} \quad V_{line} = 480$$
Then
$$I_{line} = 300,000 / (\sqrt{3} \times 480) = 361 \text{ A}$$

4. **B** F_1 represents the output of the Exclusive-Nor gate (Exclusive-OR followed by a NOT gate). We can obtain the answer by complementing the output in the truth table for Exclusive-OR gate. The function from the truth table is
$$F_1 = A_1 A_2 + \overline{A_1}\overline{A_2}$$

5. **A** If $A_1 = 1$, then $F_1 = A_2$ and if $B_1 = 1$, then $F_2 = B_2$. Therefore, $G = F_1 F_2 = A_2 B_2$.

6. **B**

7. **A** The inputs are $J = 1, K = 1$ and $D = 0$. Therefore, the binary count changes to 00.

8. **C**

9. **D**

10. **C** The closed-loop characteristic equation is given by
$$1 + \frac{K}{s(s+2)} = 0 \quad \text{or} \quad s(s+2) + K = 0$$

Therefore, K should be set to 100.

11. **A** $2\zeta\omega_n = 2$ and $\omega_n^2 = K$. If $\zeta = 1$ then $K = 1$.

12. **A** With $K = 16$ the characteristic equation is $s^2 + 2s + 16 = 0$. Therefore, $\omega_n = 4$ and $\zeta = 0.25$. The damped frequency of oscillation, which is $\omega_n\sqrt{1-\zeta^2}$, is close to 4 rad/s.

13. **B** The characteristic equation is given by
$$s(s+2)(s+10) + K = 0 \quad \text{or} \quad s^3 + 12s^2 + 20s + K = 0$$

Apply the Routh-Horwitz criterion:

$$\begin{array}{ccc} s^3 & 1 & 20 \\ s^2 & 12 & K \\ s & (240-K)/12 & 0 \\ s^0 & K & \end{array}$$

For the system to be stable, there should be no sign changes in the first column of the above table. This is possible if the value of $0 < K < 240$. The system is marginally stable if $K = 240$.

14. **A** If $K = 240$, the frequency of oscillation is obtained by solving the auxiliary equation
$$12s^2 + 240 = 0$$

which gives the frequency of oscillation of $\sqrt{20}$ rad/s or 0.71 Hz.

15. **A** Since this is a type 1 system, the steady-state error to a step input is zero and independent of gain K.

16. **D** The decimator is a time-varying operation. Therefore, the output produced by decimating the linear time-invariant (LTI) filtering will be different from the output obtained by LTI filtering followed by decimation, and the two systems are not equivalent. To see this, check the condition for time-invariance. If we shift the input to the decimator, is the output simply shifted by the same amount? In general, the answer is no. However, if the even and odd samples of the signal are identical, then shifting the input does produce a shifted output.

17. **B** The signal $x[n] = \cos(\pi n) = (-1)^n$, a sequence of alternating 1's and –1's. Hence $x[2n] = 1$ for all n, and the output of the decimator is a constant signal. The LTI filter's impulse response is $h[n] = \Delta[n] - 0.5\Delta[n-1]$, so the output of the filter in response to the constant $x[2n] = 1$ signal is also constant with value $1+0.5 = 1.5$, so $v[n] = 1.5$ for all n.

18. **A**   The convolution between the signal $x[n] = \cos(\pi n) = (-1)^n$ and filter $h[n] = \Delta[n] + \Delta[n-1]$ is zero for all time. This is easy to verify directly by computing the convolution sum. Another way to see this is to note that the z-transform of $h[n]$ has a zero at -1, and therefore this filter annihilates the frequency component at digital frequency π. Decimating the constant zero signal produces zero output, so the output of the overall system is $w[n] = 0$.

19. **D**   From the diode equation
$$i_D = I_s[e^{V_D/nV_T} - 1]$$
For Si, $n = 1$ and the thermal voltage $V_T = kT/q = 0.025\text{ V}$. When i_D reaches 90% of I_s,
$$0.9 I_s = I_s[e^{V_D/0.025} - 1]. \quad \therefore e^{V_D/0.025} = 1.9 \text{ and } V_D = 0.01605 \text{ V}$$

20. **A**   KLV for the base-emitter loop:
$$V_{BB} = V_{BE} + I_E R_E$$
$$5 = 0.7 + I_E \times 10^3. \quad \therefore I_E = 4.3 \text{ mA}$$
But
$$I_E = I_C + I_B$$
$$= \beta I_B + I_B = (\beta + 1)I_B. \quad \therefore I_B = \frac{I_E}{1+\beta} = \frac{4.3}{101} = 0.0426 \text{ mA}$$

21. **D**   Since the gate and the source are tied together, $V_{GS} = 0$:
$$I_D = I_{DSS}\left(1 - \frac{V_{GS}}{V_p}\right)^2 = I_{DSS} = 2 \text{ mA}$$
$$V_S = I_D R_S = 2 \times 10^{-3} \times 10^3 = 2 \text{ V}$$
$$V_{DS} = V_{DD} - V_S = 5 - 2 = 3 \text{ V}$$
$$V_{GD} = V_{SD} = -V_{DS} = -3 \text{ V}$$

22. **A**   We know that
$$I_D = \frac{I_{DSS}}{V_P^2}(V_{GS} - V_p)^2$$
By the voltage divider action:
$$V_{GS} = \frac{R_2}{R_1 + R_2}(-10) = \frac{10 \times 10^3}{100 \times 10^3}(-10) = -1.0 \text{ V}$$
Then
$$I_D = \frac{18}{9}(-1+3)^2 = 8 \text{ mA}$$
The voltage across the drain and source is
$$V_{SD} = 10 - I_D R_D = 10 - 8 \text{ mA} \times 1 \text{ k}\Omega = 10 - 8 = 2 \text{ V}$$

The dissipation in the JFET is

$$P_{FET} = V_{DS}I_D = 2 \times 8 \times 10^{-3} = 16 \times 10^{-3} \text{ W}$$

23. **A** The output current is

$$i_o = i_i + g_m V_{BE}$$
$$= i_i + g_m r_\pi i_i$$
$$= i_i(1 + g_m r_\pi)$$
$$= i_i(1 + \beta)$$

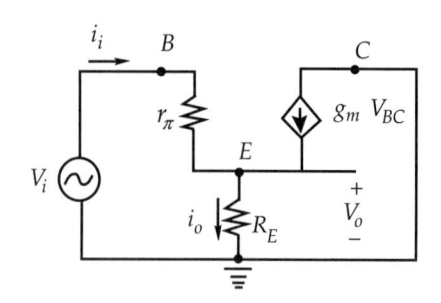

using $V_{BE} = i_i r_\pi$, $\beta = g_m r_\pi$.

small-signal (AC) equivalent circuit

Therefore, the current gain is

$$A_i = \frac{i_o}{i_i} = 1 + \beta = 1 + 100 = 101$$

24. **B** The output current is

$$i_o = i_i + g_m V_{BE}$$
$$= i_i + g_m i_i r_\pi$$
$$= i_i(1 + g_m r_\pi)$$
$$= i_i(1 + \beta)$$

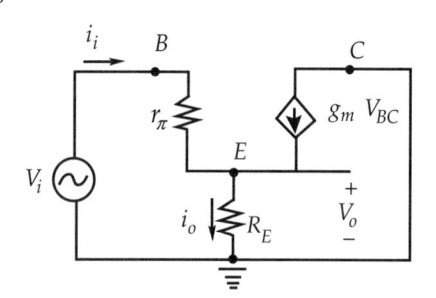

The output voltage is

$$V_o = i_o R_E$$

The input voltage is

$$V_i = i_i r_\pi + V_o$$
$$= i_i r_\pi + i_i(1+\beta)R_E = i_i[r_\pi + (1+\beta)R_E]$$

The voltage gain is

$$A_V = \frac{V_o}{V_i} = \frac{i_o R_E}{i_i[r_\pi + (1+\beta)R_E]}$$

$$= \frac{R_E}{r_\pi + (1+\beta)R_E} A_i = \frac{R_E}{r_\pi + (1+\beta)R_E}(1+\beta) = \frac{10^3(101)}{2 \times 10^3 + (101) \times 10^3} = 0.98$$

25. **D** $\frac{V_m}{V_i} = -\frac{Z_F}{Z_i}$ or $\left|\frac{V_m}{V_i}\right| = \frac{100/R_i}{\sqrt{1+(\omega^2/100)}}$

For a gain of 25 (maximum occurring at $\omega = 0$):

$$25 = \frac{100}{R_i}. \quad \therefore R_i = 4 \text{ k}\Omega$$

26. **B**   The bandwidth occurs at an amplitude of $1/\sqrt{2}$ (or 3dB in power, or 1/2 in power), which is where $\omega^2/100 = 1$ or $\omega = 10 = 2\pi f$. Therefore
$$f = 10/2\pi = 1.59 \text{ Hz}$$

27. **B**   We design the two poles so that
$$\left|\frac{V_o}{V_i}\right| = \frac{K}{\left(\sqrt{1+\omega^2/100}\right)^2} = \frac{K}{1+\omega^2/100} = \frac{K}{1+(\omega R_F C_F)^2}$$

But $\tau = R_F C_F = 10^{-1}$ which requires a parallel $R-C$ with time constant $\tau = 10^{-1}$.

28. **C** $Z = \frac{j\omega 10}{j\omega + 10} = \frac{10}{1 - j10/\omega}$

$$\left|\frac{V_o}{V_m}\right| = \frac{10/10 \times 10^3}{\sqrt{1+(10/\omega)^2}} \quad \text{and} \quad \left|\frac{V_o}{V_i}\right| = \frac{(10^5/R_i)(10/10^4)}{\sqrt{1+(10/\omega)^2}\sqrt{1+(\omega/10)^2}}$$

This results in a bandpass with
$$\left|\frac{V_o}{V_i}\right| = \frac{100/1 \text{ k}\Omega}{2} = \frac{10^{-1}}{2} = 5 \times 10^{-2}$$

centered at $\omega = 10$.

29. **A**   For V_I positive, the upper configuration acts as a voltage follower (the diode can supply current in the forward direction): $V_o = V_I$ (for $V_I > 0$). For V_I negative, the lower configuration (inverting configuration) allows $V_o/V_I = -R_2/R_1$ to be positive (and the diode can supply current in the forward direction). If $R_2 = R_1$ then for either positive or negative V_I, $V_o = |V_I|$. $\therefore V_o = |V_I| =$ full wave rectifier.

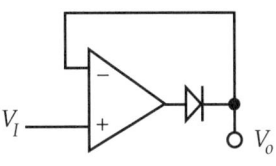

30. **B**   If R_1 is open, the upper circuit appears as a voltage follower for positive voltage and no output for negative voltage (the diode is reverse biased for V_o negative). So this results in a half-wave rectifier.

31. **D**   The input constraint (KVL) is
$$V_i = i_b R_b + 0.7$$

The output constraint (KLV) is
$$V_{ce} = V_o = -i_c R_c + 20$$

For $V_i = 1.7$

$$i_b R_b = 1.0 = i_b \times 2 \times 10^3. \quad \therefore i_b = 0.5 \text{ mA} = 5 \times 10^{-4} \text{ A}$$

Since $\beta = 100$, $i_c = \beta i_b = 5 \times 10^{-2}$ A. Then

$$V_o = -5 \times 10^{-2} \times 100 + 20 = 15 \text{ V}$$

32. **B** See No. 31. For $V_i = 3.7$

$$i_b \times 2 \times 10^3 = 3. \quad \therefore i_b = 1.5 \text{ mA} = 1.5 \times 10^{-3} \text{ A}$$

Then with $\beta = 100$, $i_c = \beta i_b = 1.5 \times 10^{-1}$ and

$$V_o = -1.5 \times 10^{-1} \times 100 + 20 = 5 \text{ V}$$

33. **C** See No. 31. For $V_i = 2.7$

$$i_b = 1.0 \text{ mA}, \; i_c = \beta i_b = 100 \text{ mA}, \text{ and } i_e = i_b + i_c = 101 \text{ mA}$$

34. **C** The dipole is created with two oppositely charged particles in close proximity. The net charge is zero. By Gauss' law:

$$\oint \varepsilon \mathbf{E} \cdot d\mathbf{A} = Q. \quad \therefore \oint \mathbf{E} \, d\mathbf{A} = 0$$

since no net charge is enclosed.

35. **D** The potential from which the dipole field is derived is

$$V = \frac{Qd \cos \theta}{4 \pi \varepsilon_o r} = \frac{K \cos \theta}{r}$$

At $r = a$ (surface of the sphere) $V = K_2 \cos \theta$.

36. **B** The dipole field satisfies Maxwell's equations:

$$\nabla \times \mathbf{E} = 0$$
$$\nabla \cdot \varepsilon \mathbf{E} = \rho \text{ since } \rho = 0$$

However at $r = 0$, two charged particles very closely spaced do exist and become a singularity. Then $\nabla \times \mathbf{E} = \nabla \cdot \mathbf{E} = 0$ everywhere except at the origin.

37. **D** The surface charges can be replaced by point charges of magnitude $4\pi a \rho_o$. The potential from the equivalent point charge at $(0,0,-a)$ is $-\left.\frac{4\pi a \rho_o}{4\pi \varepsilon_o r}\right|_{r=2a}$. The potential from $(0,0,a)$ is $\left.\frac{4\pi a \rho_o}{4\pi \varepsilon_o r}\right|_{r=a}$. Inside the sphere:

$$V = -\frac{\rho_o}{2\varepsilon_o} + \frac{\rho_o}{\varepsilon_o} = \frac{\rho_o}{2\varepsilon_o}$$

38. **A** Since equal and opposite equivalent point charges are located at $(0,0,\pm a)$, the x-y plane is at zero potential.

39. **B** The electric field due to a point charge is

$$E = \frac{Q}{4\pi\varepsilon_o r^2} \mathbf{a}_r$$

Taking care with the vector directions,

$$E = \frac{Q(-\mathbf{a}_y - \mathbf{a}_z)/\sqrt{2}}{4\pi\varepsilon_o a^2 (\sqrt{2})^2} + \frac{Q(\mathbf{a}_y - \mathbf{a}_z)/\sqrt{2}}{4\pi\varepsilon_o a^2 (\sqrt{2})^2} = \frac{-2\mathbf{a}_z(4\pi a^2 \rho_o)}{\sqrt{2}\, 8\pi\varepsilon_o a^2}$$

$$\therefore |E| = -\frac{\rho_o}{\sqrt{2}\,\varepsilon_o}$$

40. **D** At $t = 0$, $x(t)$ is a maximum, so
$$x(0) = K(2 + 4 + 2) = 8K$$

This can at most equal unity (for 100% modulation). Therefore,
$$K = \frac{1}{8} = 0.125$$

41. **C** Since
$$x_c(t) = A_c[1 + \mu x(t)]\cos 2\pi f_c t$$

The frequencies transmitted are
$$f_c \pm f_m;\quad f_c \pm 2f_m;\quad f_c \pm 3f_m$$

Since $f_c = 10 f_m$ the highest frequency carried is
$$f_c + 3f_m = 10f_m + 3f_m = 13f_m$$

42. **B** Using the signal
$$x_c(t) = A_c\left[1 + \frac{1}{8}(2\cos\omega_m t + 4\cos 2\omega_m t + 2\cos 3\omega_m t)\right]\cos 10\omega_m t$$

with $A_c = 1$, a sketch of the spectrum provides the power in the sidebands:
$$2\left[2\left(\frac{1}{16}\right)^2 + 2\left(\frac{1}{8}\right)^2 + 2\left(\frac{1}{16}\right)^2\right] = \frac{3}{32} \cong 0.1 \text{ watts}$$

43. **D** This tone-modulated signal is given by
$$x_c(t) = A_c\left[1 + \cos(6\pi \times 10^3 t)\right]\cos(600\pi \times 10^3 t)$$

The highest frequency is then 3 kHz + 300 kHz = 303 kHz

44. **A** The power dissipated can be obtained by considering the positive spectrum (phasor representation) of
$$x_c(t) = A_c \cos(600\pi \times 10^3 t) + \frac{A_c}{2}\cos(606\pi \times 10^3 t) + \frac{A_c}{2}\cos(594\pi \times 10^3 t)$$

The power dissipated is then

$$\langle P \rangle = \frac{1}{2}\left[A_c^2 + \left(\frac{A_c}{2}\right)^2 \times 2\right] = \frac{3}{4}A_c^2$$

45. A The USSB-SC for the positive spectrum (phasor representation) is given by

$$x_{USSC-SC} = \frac{A_c}{2}\cos(606\pi \times 10^3 t)$$

46. C Since the current through an inductor cannot change instantaneously (unless an infinite voltage can be supported),

$$i(0^-) = i(0^+) = 0$$

47. B Since the current at $t = 0^+$ is initially zero, all the voltage V_s must initially drop across the inductor. But

$$v_L = L\frac{di}{dt} = V_s \text{ at } t = 0^+. \quad \therefore \frac{V_s}{L} = i'(0^+)$$

48. B The roots for the RLC circuit (characteristic equation)

$$s_{1,2} = -\frac{R}{2L} \pm \sqrt{\left(\frac{R}{2L}\right)^2 - \frac{1}{LC}}$$

give overdamped (exponential decay) when

$$\left(\frac{R}{2L}\right)^2 > \frac{1}{LC}$$

49. C Since $i(0^+) = 0$ then $\phi = \frac{\pi}{2}$. Note: this could also be $\phi = -\frac{\pi}{2}$. Since current starts at zero at $t = 0$ it builds up as a sine, but since K was not specified it could be negative.

50. C If the system is linear, which is not given, both a Norton and Thevenin are possible. Since it is not possible to evaluate nonlinearity by two measurements only answer **C** is acceptable.

51. D Given that the system is linear, a Thevenin is possible; this satisfies (A). Given the two measurements:

$$v_T = R_T\left(\frac{20}{2}\right) + 20$$

$$v_T = R_T\left(\frac{25}{5}\right) + 25$$

A Thevenin with $v_T = 30$ V and $R_T = 1$ kΩ results. A 4 kΩ resistor would result in power dissipated of

$$P = i^2 R = 36 \times 10^{-6} \times 4 \times 10^3 = 0.144 \text{ W}$$

Thus (B) is also acceptable. Since the system is linear a Norton is possible; this satisfies (C). Consequently (D) is the answer.

52. **C** Since the stack pointer is decremented on a load, it will decrease in value and hence the stack will fill up toward lower memory addresses.

53. **D** The SP will be decremented 10 times and hence be $0000_{16} - A_{16} = FFF6_{16}$.

54. **C** A stack is not effective for random access of data.

55. **D** The flip-flop icon indicates edge-sensitivity with the ∧ symbol and falling-edge with the O symbol.

56. **A** The flip-flop has the following next state table:

A	B	Q_n	Q_{n+1}
0	0	0	1
0	0	1	0
0	1	0	1
0	1	1	1
1	0	0	0
1	0	1	0
1	1	0	0
1	1	1	1

From this, a K-map can be formed to determine the characteristic equation:

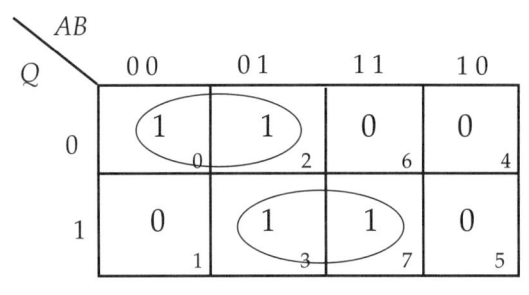

$$Q_{n+1} = \overline{A}\overline{Q}_n + BQ_n$$

57. **B** This is found by using the truth table in the problem statement for evaluating all conditions on A and B which force the desired transition from Q_n to Q_{n+1}.

58. **A**

59. **C**

60. **D**

Equation Summaries

Appendix A

The following pages are Equation Summary Sheets of the FE/EIT subjects which rely most on equations. They are intended to be used for quick overviews, handy problem solving—and as part of a special strategy for preparing for the exam in its new format.

The selected equations are presented in the same format and nomenclature as is to be found in the NCEES Reference Handbook, a newsprint booklet which is given to all exam applicants. Be prepared! You may find some of the nomenclature to be different from what you are used to using, as the NCEES has apparently obtained some of it from older or obscure texts.

The current edition of the NCEES Reference Handbook has many extraneous equations which will in all likelihood be of little aid in solving exam problems. As discussed earlier, an excellent exam preparation strategy is to identify the equations you find to be most useful, then highlight them and become acquainted with their position in the Handbook, so that you may quickly access them in the "clean" Handbook during the exam itself. This strategy should maximize your ability to perform as well as possible, given the present exam format.

A programmable calculator may be pre-programmed to solve many of the problems on the FE exam, e.g., problems involving matrices, permutations, standard deviations, Mohr's circle problems, etc. Or you may choose to buy the HP-48G, which has hundreds of equations and constants preprogrammed. (We offer this superior calculator at a significant discount.) Not all states allow the use of the HP-48G, so be sure you check with your state board.

These summaries are useful for study

They can help you anticipate and prepare for NCEES obstacles

Calculators are a good way to access equations during the exam

Mathematics
—Selected Equations from the NCEES Reference Handbook—

Straight Line: $y = mx + b$ (slope - intercept form) $m = \dfrac{y_2 - y_1}{x_2 - x_1}$ (slope)

$y - y_1 = m(x - x_1)$ (point - slope form) $m_1 = -\dfrac{1}{m_2}$ (two perpendicular lines)

Quadratic Equation: $ax^2 + bx + c = 0$ $\text{roots} = \dfrac{-b \pm \sqrt{b^2 - 4ac}}{2a}$

Conic Sections:

	General Form	$h = k = 0$
Parabola:	$(y - k)^2 = 2p(x - h)$	$y^2 = 2px$ Focus: $(p/2, 0)$ Directrix: $x = -p/2$
Ellipse:	$\dfrac{(x-h)^2}{a^2} + \dfrac{(y-k)^2}{b^2} = 1$	$\dfrac{x^2}{a^2} + \dfrac{y^2}{b^2} = 1$ Focus: $\left(\sqrt{a^2 - b^2},\, 0\right)$
Hyperbola:	$\dfrac{(x-h)^2}{a^2} - \dfrac{(y-k)^2}{b^2} = 1$	$\dfrac{x^2}{a^2} - \dfrac{y^2}{b^2} = 1$ Focus: $\left(\sqrt{a^2 + b^2},\, 0\right)$
Circle:	$(x - h)^2 + (y - k)^2 = r^2$	$x^2 + y^2 = r^2$

Logarithms:

$\ln x = 2.3026 \log x$ $\log xy = \log x + \log y$ $\log x/y = \log x - \log y$

$\log_b b^n = n$ $\log_b b = 1$

$\log x^c = c \log x$ $\log 1 = 0$ If $b^c = x$, then $\log_b x = c$

Trigonometry:

$\sin \theta = y/r$ $\cos \theta = x/r$

$\tan \theta = y/x$ $\cot \theta = x/y$

$\csc \theta = r/y$ $\sec \theta = r/x$

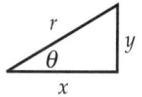

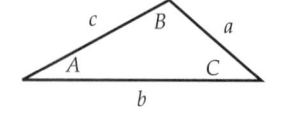

Law of Sines: $\dfrac{a}{\sin A} = \dfrac{b}{\sin B} = \dfrac{c}{\sin C}$

Law of Cosines: $a^2 = b^2 + c^2 - 2bc \cos A$

Identities:

$\tan \theta = \sin \theta / \cos \theta$ $\sin 2\alpha = 2 \sin \alpha \cos \alpha$

$\sin^2 \theta + \cos^2 \theta = 1$ $\cos 2\alpha = \cos^2 \alpha - \sin^2 \alpha$

$\sin(\alpha + \beta) = \sin \alpha \cos \beta + \cos \alpha \sin \beta$ $= 2 \cos^2 \alpha - 1$

$\cos(\alpha + \beta) = \cos \alpha \cos \beta - \sin \alpha \sin \beta$ $= 1 - 2 \sin^2 \alpha$

Complex Numbers:
$$i = \sqrt{-1} \qquad x + iy = re^{i\theta} \qquad \cos\theta = \frac{e^{i\theta} + e^{-i\theta}}{2}$$

$$r = \sqrt{x^2 + y^2} \qquad e^{i\theta} = \cos\theta + i\sin\theta$$

$$\sin\theta = \frac{e^{i\theta} - e^{-i\theta}}{2i}$$

$$(x + iy)^n = r^n(\cos n\theta + i\sin n\theta)$$

Matrices:

Transpose: $\mathbf{B} = \mathbf{A}^T$ if $b_{ji} = a_{ij}$

Inverse: $\mathbf{A}^{-1} = \dfrac{\text{adj}(\mathbf{A})}{|\mathbf{A}|}$

Adjoint: $\text{adj}(\mathbf{A}) =$ matrix formed by replacing $\mathbf{A}^T$ elements with their cofactors

Cofactor: cofactor = minor $\times (-1)^{h+k}$ where h = column, k = row

Minor: minor = determinant that remains after the common row and column are struck out

Vectors:
$$\mathbf{A} \cdot \mathbf{B} = a_x b_x + a_y b_y + a_z b_z \qquad \mathbf{A} \times \mathbf{B} = \begin{vmatrix} \mathbf{i} & \mathbf{j} & \mathbf{k} \\ a_x & a_y & a_z \\ b_x & b_y & b_z \end{vmatrix}$$

$$= |\mathbf{A}||\mathbf{B}|\cos\theta = \mathbf{B} \cdot \mathbf{A}$$

$$= |\mathbf{A}||\mathbf{B}|\mathbf{n}\sin\theta = -\mathbf{B} \times \mathbf{A} \quad \text{where } \mathbf{n} \text{ is } \perp \text{ plane of } \mathbf{A} \text{ and } \mathbf{B}$$

$$\mathbf{i} \cdot \mathbf{i} = \mathbf{j} \cdot \mathbf{j} = \mathbf{k} \cdot \mathbf{k} = 1 \qquad \mathbf{i} \times \mathbf{j} = \mathbf{k}, \quad \mathbf{j} \times \mathbf{k} = \mathbf{i}, \quad \mathbf{k} \times \mathbf{i} = \mathbf{j}$$

Taylor Series:
$$f(x) = f(a) + \frac{f'(a)}{1!}(x-a) + \frac{f''(a)}{2!}(x-a)^2 + \cdots$$

Maclaurin Series: a Taylor series with $a = 0$

Probability and Statistics:

$$P(n,r) = \frac{n!}{(n-r)!} \qquad \text{(permutation of } n \text{ things taken } r \text{ at a time)}$$

$$C(n,r) = \frac{P(n,r)}{r!} = \frac{n!}{r!(n-r)!} \qquad \text{(combination of } n \text{ things taken } r \text{ at a time)}$$

$$\bar{x} = \frac{x_1 + x_2 + \cdots + x_n}{n} \qquad \text{(arithmetic mean)}$$

$$\sigma^2 = \frac{\sum(x_i - \bar{x})^2}{n-1} \qquad \text{(variance)}$$

$$\sigma = \sqrt{\text{variance}} \qquad \text{(sample standard deviation)}$$

$$\text{median} = \begin{cases} \text{middle value if odd number of items} \\ \frac{1}{2}(\text{sum of middle two values}) \text{ if even number of items} \end{cases}$$

mode = value that occurs most often

Appendix A / Equation Summaries—*Mathematics*

Calculus: $f'(x) = 0 \begin{cases} \text{maximum} & \text{if } f''(x) < 0 \\ \text{minimum} & \text{if } f''(x) > 0 \end{cases}$

L'Hospital's Rule: $\lim\limits_{x \to a} \dfrac{f(x)}{g(x)} = \lim\limits_{x \to a} \dfrac{f'(x)}{g'(x)}$ if $\dfrac{f(a)}{g(a)} = \dfrac{0}{0}$ or $\dfrac{\infty}{\infty}$

$$\frac{d}{dx}(uv) = u\frac{dv}{dx} + v\frac{du}{dx} \qquad \frac{d}{dx}(\ln u) = \frac{1}{u}\frac{du}{dx} \qquad \frac{d}{dx}(\sin u) = \cos u \frac{du}{dx}$$

$$\frac{d}{dx}\left(\frac{u}{v}\right) = \frac{v\,du/dx - u\,dv/dx}{v^2} \qquad \frac{d}{dx}(e^u) = e^u \frac{du}{dx} \qquad \frac{d}{dx}(\cos u) = -\sin u \frac{du}{dx}$$

$$\frac{d}{dx}(u^n) = nu^{n-1}\frac{du}{dx}$$

$$\int x^n dx = \frac{x^{n+1}}{n+1} \quad n \neq -1 \qquad \int \sin x\, dx = -\cos x \qquad \int \sin^2 x\, dx = \frac{x}{2} - \frac{\sin 2x}{4}$$

$$\int \frac{dx}{ax+b} = \frac{1}{a}\ln|ax+b| \qquad \int \cos x\, dx = \sin x \qquad \int \cos^2 x\, dx = \frac{x}{2} + \frac{\sin 2x}{4}$$

$$\int e^{ax} dx = \frac{1}{a}e^{ax}$$

Differential Equations: $y'' + 2ay' + by = f(x)$ (linear, 2nd order, constant coefficient, nonhomogeneous)

Homogeneous solution: $y_h(x) = C_1 e^{r_1 x} + C_2 e^{r_2 x}$ if $r_1 \neq r_2$ where $r^2 + 2ar + b = 0$

$\qquad\qquad\qquad\qquad\quad = (C_1 + C_2 x)e^{r_1 x}$ if $r_1 = r_2$

$\qquad\qquad\qquad\qquad\quad = e^{-ax}(C_1 \cos \beta x + C_2 \sin \beta x)$ if $a^2 < b.$ $\beta = \sqrt{b-a^2}$

Particular solution: $y_p = B$ if $f(x) = A$

$\qquad\qquad\qquad\quad = Be^{\alpha x}$ if $f(x) = Ae^{\alpha x}$

$\qquad\qquad\qquad\quad = B_1 \sin \omega x + B_2 \cos \omega x$ if $f(x) = A_1 \sin \omega x + A_2 \cos \omega x$

General solution: $y(x) = y_h(x) + y_p(x)$

Mechanics of Materials
—Selected equations from the NCEES Reference Handbook—

Definitions:
$\sigma = \varepsilon E$
$\tau = \gamma G$
$E = 2G(1 + v)$
$v = -\dfrac{\varepsilon_{lateral}}{\varepsilon_{longitudinal}}$

E = modulus of elasticity
G = shear modulus
σ and τ = normal and shear stress
ε and γ = normal and shear strain
v = Poisson's ratio

Uniaxial Loading:
$\left.\begin{array}{l}\sigma = \dfrac{P}{A} \\ \varepsilon = \dfrac{\delta}{L}\end{array}\right\} \quad \delta = \dfrac{PL}{AE}$

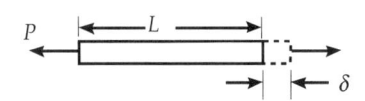

Thermal Deformation:
$\delta_t = \alpha L(T - T_o)$
α = coefficient of thermal expansion

Thin-walled Pressure Vessel:
$\sigma_t = \dfrac{pD}{2t}$ hoop (circumferential) stress t = cylinder thickness
$\sigma_a = \dfrac{pD}{4t}$ axial (longitudinal) stress D = cylinder diameter
                                                                                                   p = pressure

Stress and Strain:

Stress Condition   *Mohr's Circle*   *Maximum and Minimum Stresses*

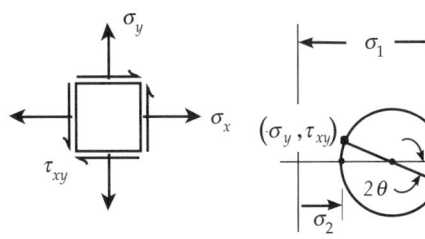

$\sigma_1 = \sigma_{max} = \dfrac{\sigma_x + \sigma_y}{2} + \left[(\sigma_x - \sigma_y)^2/4 + \tau_{xy}^2\right]^{1/2}$

$\sigma_2 = \sigma_{min} = \dfrac{\sigma_x + \sigma_y}{2} - \left[(\sigma_x - \sigma_y)^2/4 + \tau_{xy}^2\right]^{1/2}$

$\tau_{max} = \dfrac{\sigma_1 - \sigma_2}{2}$ = radius of Mohr's circle

3-D Strain: $\varepsilon_x = \dfrac{1}{E}\left[\sigma_x - v(\sigma_y + \sigma_z)\right]$
$\gamma_{xy} = \dfrac{\tau_{xy}}{G}$

Torsion:
$\tau = \dfrac{Tr}{J}$ (shear stress) J = polar moment of inertia
$\phi = \dfrac{TL}{JG}$ (angle of twist) $= \pi r^4/2$ for a circle

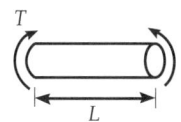

Beams:
$V = \dfrac{dM}{dx}$ V = vertical shear force M = bending moment

$\sigma = -\dfrac{My}{I}$ I = centroidal moment of inertia
                $= bh^3/12$ for a rectangle y = distance from neutral axis
                $= \pi r^4/4$ for a circle

$\tau = \dfrac{VQ}{Ib}$ Q = moment of area between y-position and top or bottom

$EIy'' = M$ differential equation of deflection curve

Columns: $P_{cr} = \dfrac{\pi^2 EI}{k^2 L^2}$ $k = \begin{cases} 1 & \text{ends pinned} \\ 0.5 & \text{ends fixed} \\ 0.7 & \text{one pinned, one fixed} \\ 2 & \text{one fixed, one free} \end{cases}$

Dynamics
—Selected equations from the NCEES Reference Handbook—

Kinematics (motion only)

Tangential and Normal Components:

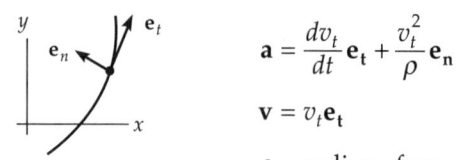

$$\mathbf{a} = \frac{dv_t}{dt}\mathbf{e_t} + \frac{v_t^2}{\rho}\mathbf{e_n}$$

$$\mathbf{v} = v_t\mathbf{e_t}$$

ρ = radius of curvature

Plane Circular Motion:

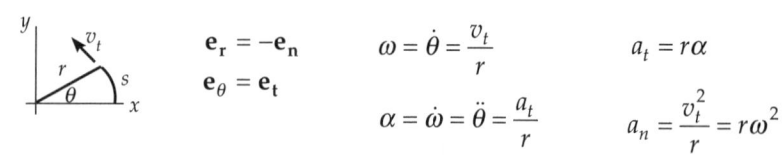

$\mathbf{e_r} = -\mathbf{e_n}$

$\mathbf{e_\theta} = \mathbf{e_t}$

$\omega = \dot{\theta} = \frac{v_t}{r}$

$\alpha = \dot{\omega} = \ddot{\theta} = \frac{a_t}{r}$

$v_t = r\omega$

$a_t = r\alpha$

$a_n = \frac{v_t^2}{r} = r\omega^2$

$s = r\theta$

Straight Line Motion:

$s = s_o + v_o t + a_o t^2/2$

$v = v_o + a_o t$

$v^2 = v_o^2 + 2a_o(s - s_o)$

Projectile Motion:

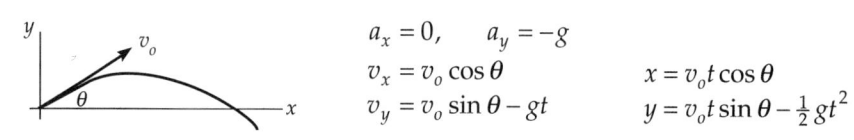

$a_x = 0, \quad a_y = -g$

$v_x = v_o \cos\theta$

$v_y = v_o \sin\theta - gt$

$x = v_o t \cos\theta$

$y = v_o t \sin\theta - \frac{1}{2}gt^2$

Kinematics (forces and motion)

$$\sum \mathbf{F} = \frac{d}{dt}(m\mathbf{v}), \quad \sum F_t = ma_t = m\frac{dv_t}{dt}, \quad \sum F_n = ma_n = m\frac{v_t^2}{\rho}$$

Impulse and Momentum:

$$m[v_x(t) - v_x(0)] = \int_0^t F_x(t)dt \quad \text{or} \quad \text{change in momentum = impulse}$$

Work and Energy: $PE_1 + KE_1 + W_{1\to 2} = PE_2 + KE_2$ where

$KE = \frac{1}{2}mv^2$

$PE = mgh$ (gravity)

$\quad = \frac{1}{2}kx^2$ (spring)

$W_{1\to 2}$ = friction force work

Impact: $m_1 v_1 + m_2 v_2 = m_1 v_1' + m_2 v_2'$

$$e = -\frac{v_{1n}' - v_{2n}'}{v_{1n} - v_{2n}} = \begin{cases} 1 & \text{elastic} \\ 0 & \text{plastic} \end{cases}$$

v_1, v_2 = velocities before impact

v_1', v_2' = velocities after impact

Rotation: $I_o \alpha = \sum M_o$ where $I_o = \int (x^2 + y^2) dm$, rotation about O.

constant M: $\alpha = \frac{M}{I}$

$\omega = \omega_o + \frac{M}{I} t$

$\theta = \theta_o + \omega_o t + \frac{M}{2I} t^2$

work and energy: $I_o \frac{\omega^2}{2} - I_o \frac{\omega_o^2}{2} = \int_{\theta_o}^{\theta} M d\theta$

Banking of Curves: $\tan\theta = \frac{v^2}{rg}$ where r = radius of curvature

θ = angle between surface and horizontal

Electric Circuits
—Selected Equations from the NCEES Reference Handbook—

Electrostatics:

$F_2 = \dfrac{Q_1 Q_2}{4\pi \varepsilon r^2}$ (force on charge 2 due to charge 1) $\quad \varepsilon =$ permittivity — $C^2/N \cdot m^2 = F/m$
$\hspace{9cm} = 8.85 \times 10^{-12}$ for air or free space

$E = \dfrac{Q}{4\pi \varepsilon r^2}$ (electric field intensity due to point charge Q — C)

$E_L = \dfrac{\rho_L}{2\pi \varepsilon r}$ (radial field due to line charge ρ_L — C/m)

$E_s = \dfrac{\rho_s}{2\varepsilon}$ (plane field due to sheet charge ρ_s — C/m^2)

$Q = \oint \varepsilon \mathbf{E} \cdot d\mathbf{A}$ (enclosed charge — C)

$E = \dfrac{V}{d}$ (electric field between plates with potential difference V separated by the distance d)

$H = \dfrac{I}{2\pi r}$ (magnetic field strength due to current in long wire)

$B = \mu H$ (magnetic flux density)

$\mathbf{F} = I\mathbf{L} \times \mathbf{B}$ (force on conductor) $\quad\quad \mathbf{L} =$ length vector of conductor

DC Circuits:

Resistors: $\quad V = IR$ (Ohm's law) $\quad R_T = R_1 + R_2 + \cdots$ (series)

$P = VI = \dfrac{V^2}{R} = I^2 R$ (power) $\quad R_T = \left[\dfrac{1}{R_1} + \dfrac{1}{R_2} + \cdots\right]^{-1}$ (parallel)

Capacitors: $\quad i = C\dfrac{dv}{dt} \quad$ energy stored $= \tfrac{1}{2} C v^2 \quad C_{eq} = C_1 + C_2 + \cdots$ (parallel)

$v = \dfrac{1}{C} \int i\, dt \quad\quad C_{eq} = \left[\dfrac{1}{C_1} + \dfrac{1}{C_2} + \cdots\right]^{-1}$ (series)

Inductors: $\quad i = \dfrac{1}{L} \int v\, dt \quad$ energy stored $= \tfrac{1}{2} L i^2 \quad L_{eq} = L_1 + L_2 + \cdots$ (series)

$v = L\dfrac{di}{dt} \quad\quad L_{eq} = \left[\dfrac{1}{L_1} + \dfrac{1}{L_2} + \cdots\right]^{-1}$ (parallel)

Kirchhoff Voltage Law (KVL): $\quad \sum V_{rises} = \sum V_{drops} = 0$

Kirchhoff Current Law (KCL): $\quad \sum I_{in} = \sum I_{out}$

Thevenin equivalent circuit:

$R_{eq} = \dfrac{V_{eq}}{I_{sc}} \quad\quad I_{sc} =$ short circuit current
$\hspace{2.3cm} V_{eq} =$ open circuit voltage

RC Transients:

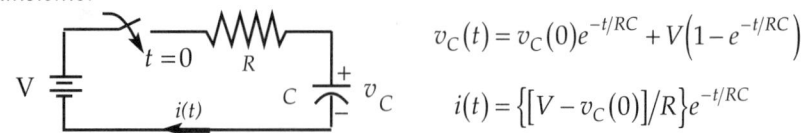

$$v_C(t) = v_C(0)e^{-t/RC} + V(1-e^{-t/RC})$$

$$i(t) = \{[V - v_C(0)]/R\}e^{-t/RC}$$

RL Transients:

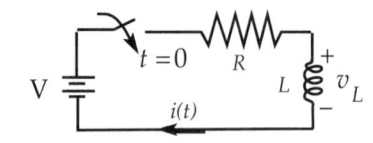

$$v_L(t) = -i(0)Re^{-Rt/L} + Ve^{-Rt/L}$$

$$i(t) = i(0)e^{-Rt/L} + \frac{V}{R}(1-e^{-Rt/L})$$

Operational Amplifiers:

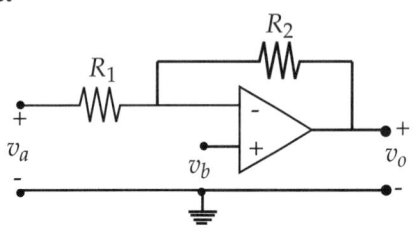

$$v_o = -\frac{R_2}{R_1}v_a + \left(1 + \frac{R_2}{R_1}\right)v_b$$

inverting if $v_b = 0$
non-inverting if $v_a = 0$

AC Circuits: (single phase)

$$f = \frac{1}{T} = \frac{\omega}{2\pi}$$

f = frequency (Hz)
T = period (sec)
ω = angular frequency (rad/s)

$$V_{avg} = \frac{2}{\pi}V_{max} \quad \text{(full-wave rectified sine wave)}$$

$$V_{avg} = \frac{1}{\pi}V_{max} \quad \text{(half-wave rectified sine wave)}$$

$$V_{rms} = \frac{1}{\sqrt{2}}V_{max} \quad \text{(full-wave rectified sine wave)}$$

$$V_{rms} = \frac{1}{2}V_{max} \quad \text{(half-wave rectified sine wave)}$$

Resistor: $Z = R$ $\quad Z$ = Impedance

Capacitor: $Z = -\dfrac{j}{\omega C} = -jX$ $\quad X$ = Reactance

Inductor: $Z = j\omega L = jX$

$V = IZ$

$P = \frac{1}{2}V_{max}I_{max}\cos\theta = V_{rms}I_{rms}\cos\theta$ (real power) ($\theta = 0$ for resistors)

$Q = \frac{1}{2}V_{max}I_{max}\sin\theta = V_{rms}I_{rms}\sin\theta$ (reactive power)

p.f. = $\cos\theta$ (power factor)

Resonance: $f = \dfrac{1}{2\pi\sqrt{LC}}$ (resonant frequency for series and parallel circuits)

Fluid Mechanics
—Selected Equations from the NCEES Reference Handbook—

Properties: $\rho = \dfrac{m}{V}$ (density) $\qquad \tau_n = -p$ (normal stress)

$\gamma = \rho g$ (specific weight) $\qquad \tau_t = \mu \dfrac{dv}{dy}$ (tangential stress)

$v = \dfrac{\mu}{\rho}$ (kinematic viscosity) $\qquad \mu =$ dynamic viscosity

Statics: $p_2 - p_1 = -\gamma h$ (h is vertical upward) $\qquad F = \gamma h_C A$

$F_{\text{buoyant}} = \gamma V_{\text{displaced}}$ (Archimedes' principle) $\qquad z^* = \dfrac{I_C}{A Z_C}$

One-Dimensional Flows: $A_1 V_1 = A_2 V_2$ (continuity equation)

$Q = AV$ (flow rate)

$\dot{m} = \rho A V$ (mass flow rate)

$-\dfrac{\dot{W}_s}{\gamma Q} + \dfrac{p_1}{\gamma} + \dfrac{V_1^2}{2g} + z_1 = \dfrac{p_2}{\gamma} + \dfrac{V_2^2}{2g} + z_2 + h_f$ (Energy Equation — if $h_f = \dot{W}_s = 0$, then Bernoulli Eq.)

$h_f = f \dfrac{L}{D} \dfrac{V^2}{2g}$ (Darcy's Equation — find f on Moody Diagram)

$\text{Re} = \dfrac{VD\rho}{\mu}$ (Reynolds Number)

$h_{f,\text{ fitting}} = C \dfrac{V^2}{2g}$ (minor losses — C is loss coefficient)

$\sum \mathbf{F} = \rho Q (\mathbf{V}_2 - \mathbf{V}_1)$ (Momentum equation)

Perfect Gas: $p = \rho RT$ (perfect gas law)

$c = \sqrt{kRT}$ (speed of sound)

$\text{M} = \dfrac{V}{c}$ (Mach number)

Similitude: If viscous effects dominate (internal flows) then Reynolds numbers on prototype and model are equated:

$(\text{Re})_p = (\text{Re})_m \quad \text{or} \quad \left(\dfrac{V\ell\rho}{\mu}\right)_p = \left(\dfrac{V\ell\rho}{\mu}\right)_m$

If gravity dominates (dams, weirs, ships) then Froude numbers are equated:

$(\text{Fr})_p = (\text{Fr})_m \quad \text{or} \quad \left(\dfrac{V^2}{\ell g}\right)_p = \left(\dfrac{V^2}{\ell g}\right)_m$

Open Channel: $Q = \dfrac{C}{n} A R^{2/3} S^{1/2}$ where $R = \dfrac{A}{P_{\text{wetted}}}$

$C = \begin{cases} 1.0 & \text{metric} \\ 1.49 & \text{english} \end{cases}$

Thermodynamics
—Selected Equations from the NCEES Reference Handbook—

Properties:
- P (absolute pressure, kPa or lbf/in^2)
- $v = \dfrac{V}{m}$ (specific volume, m^3/kg or ft^3/lbm)
- u (internal energy, kJ/kg or Btu/lbm)
- $h = u + Pv$ (enthalpy, kJ/kg or Btu/lbm)
- s (entropy, kJ/kg·K or Btu/lbm-°R)
- c_p (constant pressure specific heat, kJ/kg·K or Btu/lbm-°R)
- c_v (constant volume specific heat, kJ/kg·K or Btu/lbm-°R)
- $x = \dfrac{m_v}{m_{total}}$ (quality)

Two phase system:
$$v = v_f + xv_{fg} \quad \text{where} \quad v_{fg} = v_g - v_f$$
$$h = h_f + xh_{fg}$$
v_f = saturated liquid value
v_g = saturated vapor value

Ideal gas:
$$Pv = RT, \quad PV = mRT \quad \text{where } R = \dfrac{\overline{R}}{M}, \quad \overline{R} = 8.314 \dfrac{\text{kJ}}{\text{kmol} \cdot \text{K}} \text{ or } 1545 \dfrac{\text{ft-lbf}}{\text{lbmol-°R}}$$

$$\Delta u = c_v \Delta T, \quad \Delta h = c_p \Delta T$$

$$\Delta s = c_p \ln \dfrac{T_2}{T_1} - R \ln \dfrac{P_2}{P_1} = c_v \ln \dfrac{T_2}{T_1} + R \ln \dfrac{v_2}{v_1}$$

$$\left. \dfrac{T_2}{T_1} = \left(\dfrac{P_2}{P_1}\right)^{\frac{k-1}{k}} = \left(\dfrac{v_1}{v_2}\right)^{k-1}, \quad \begin{array}{l} P_2 v_2^k = P_1 v_1^k \\ k = c_p/c_v \end{array} \right\} \text{(constant entropy process)}$$

First law (system): $q - w = \Delta u \quad$ where $\quad w = \int P dv$

$$= RT \ln \dfrac{v_2}{v_1} = RT \ln \dfrac{P_1}{P_2} \quad \text{(isothermal process with ideal gas)}$$

First law (control volume):
- $h_i + V_i^2/2 = h_e + V_e^2/2$ (nozzles, diffusers)
- $h_i = h_e + w$ (turbine, compressor)
- $h_i = h_e$ (throttling device, valve)
- $h_i + q = h_e$ (boilers, condensers, evaporators)

i = inlet
e = exit

Cycles: $\eta = \dfrac{W}{Q_H} = \dfrac{Q_H - Q_L}{Q_H}$ (efficiency) $\quad$ COP $= \dfrac{Q_H}{W}$ (heat pump)

$\qquad\qquad = 1 - \dfrac{T_L}{T_H}$ (Carnot cycle) $\quad\qquad\qquad = \dfrac{Q_L}{W}$ (refrigerator)

Second Law: No engine can produce work while transferring heat with a single reservoir. (Kelvin-Planck)

No refrigerator can operate without a work input. (Clausius)

$\Delta S \geq \int \dfrac{\delta Q}{T} \qquad\qquad \Delta S = \dfrac{Q}{T}$ (reservoir or T = const)

$\Delta S_{total} = \Delta S_{surr} + \Delta S_{system} \geq 0 \qquad \Delta S = C_p \ln \dfrac{T_2}{T_1}$ (solid or liquid)

Heat Transfer: $q = -kA \dfrac{dT}{dx}$ (conduction) $\qquad k$ = conductivity

$\qquad\qquad\quad = -kA \dfrac{T_2 - T_1}{L}$ (through a wall) $\qquad R = \dfrac{L}{kA}$ (resistance factor)

$q = hA(T_1 - T_2)$ (convection) $\qquad\qquad R = \dfrac{1}{hA}$ (resistance factor)

$\quad = \varepsilon \sigma A \left(T_1^4 - T_2^4 \right) F_{12}$ (radiation) $\qquad h$ = convection coefficient

$\varepsilon = 1$ for black body (emissivity)

$\sigma = 5.67 \times 10^{-8} \dfrac{W}{m^2 \cdot K^4}$ (Stefan - Boltzmann constant)

$F_{12} = 1$ if one body encloses the other (shape factor)

English and SI Units

Appendix B

The following tables present the SI (Systems International) units and the conversion of English units to SI units, along with some of the more common conversion factors.

SI Prefixes

Multiplication Factor	Prefix	Symbol
10^{15}	peta	P
10^{12}	tera	T
10^{9}	giga	G
10^{6}	mega	M
10^{3}	kilo	k
10^{-1}	deci	d
10^{-2}	centi	c
10^{-3}	mili	m
10^{-6}	micro	µ
10^{-9}	nano	n
10^{-12}	pico	p
10^{-15}	femto	f

SI Base Units

Quantity	Name	Symbol
length	meter	m
mass	kilogram	kg
time	second	s
electric current	ampere	A
temperature	kelvin	K
amount of substance	mole	mol
luminous intensity	candela	cd

SI Derived Units

Quantity	Name	Symbol	In Terms of Other Units
area	square meter		m^2
volume	cubic meter		m^3
velocity	meter per second		m/s
acceleration	meter per second squared		m/s^2
density	kilogram per cubic meter		kg/m^3
specific volume	cubic meter per kilogram		m^3/kg
frequency	hertz	Hz	s^{-1}
force	newton	N	$m \cdot kg/s^2$
pressure, stress	pascal	Pa	$kg/(m \cdot s^2)$
energy, work, heat	joule	J	$N \cdot m$
power	watt	W	J/s
electric charge	coulomb	C	$A \cdot s$
electric potential	volt	V	W/A
capacitance	farad	F	C/V
electric resistance	ohm	Ω	V/A
conductance	siemens	S	A/V
magnetic flux	weber	Wb	$V \cdot s$
inductance	henry	H	Wb/A
viscosity	pascal second		$Pa \cdot s$
moment (torque)	meter newton		$N \cdot m$
heat flux	watt per square meter		W/m^2
entropy	joule per kelvin		J/K
specific heat	joule per kilogram-kelvin		$J/(kg \cdot K)$
conductivity	watt per meter-kelvin		$W/(m \cdot K)$

Conversion Factors to SI Units

English	SI	SI Symbol	To Convert from English to SI Multiply by
Area			
square inch	square centimeter	cm2	6.452
square foot	square meter	m2	0.09290
acre	hectare	ha	0.4047
Length			
inch	centimeter	cm	2.54
foot	meter	m	0.3048
mile	kilometer	km	1.6093
Volume			
cubic inch	cubic centimeter	cm^3	16.387
cubic foot	cubic meter	m^3	0.02832
gallon	cubic meter	m^3	0.003785
gallon	liter	L	3.785
Mass			
pound mass	kilogram	kg	0.4536
slug	kilogram	kg	14.59
Force			
pound	newton	N	4.448
kip(1000 lb)	newton	N	4448
Density			
pound/cubic foot	kilogram/cubic meter	kg/m^3	16.02
pound/cubic foot	grams/liter	g/L	16.02
Work, Energy, Heat			
foot-pound	joule	J	1.356
Btu	joule	J	1055
Btu	kilowatt-hour	kWh	0.000293
therm	kilowatt-hour	kWh	29.3

Conversion Factors to SI Units (continued)

English	SI	SI Symbol	To Convert from English to SI Multiply by
Power, Heat, Rate			
horsepower	watt	W	745.7
foot pound/sec	watt	W	1.356
Btu/hour	watt	W	0.2931
Btu/hour-ft^2-°F	watt/meter squared-°C	W/m$^2 \cdot$°C	5.678
tons of refrig.	kilowatts	kW	3.517
Pressure			
pound/square inch	kilopascal	kPa	6.895
pound/square foot	kilopascal	kPa	0.04788
inches of H$_2$O	kilopascal	kPa	0.2486
inches of Hg	kilopascal	kPa	3.374
one atmosphere	kilopascal	kPa	101.3
Temperature			
Fahrenheit	Celsius	°C	5 (°F − 32)/9
Fahrenheit	kelvin	K	5 (°F + 460)/9
Velocity			
foot/second	meter/second	m/s	0.3048
mile/hour	meter/second	m/s	0.4470
mile/hour	kilometer/hour	km/h	1.609
Acceleration			
foot/second squared	meter/second squared	m/s^2	0.3048
Torque			
pound-foot	newton-meter	N$\cdot$m	1.356
pound-inch	newton-meter	N$\cdot$m	0.1130
Viscosity, Kinematic Viscosity			
pound-sec/square foot	newton-sec/square meter	N$\cdot$s/m^2	47.88
square foot/second	square meter/second	m^2/s	0.09290
Flow Rate			
cubic foot/minute	cubic meter/second	m^3/s	0.0004719
cubic foot/minute	liter/second	L/s	0.4719
Frequency			
cycles/second	hertz	Hz	1.00

Conversion Factors

Length
1 cm = 0.3937 in
1 m = 3.281 ft
1 yd = 3 ft
1 mi = 5280 ft
1 mi = 1760 yd
1 km = 3281 ft

Area
1 cm^2 = 0.155 in^2
1 m^2 = 10.76 ft^2
1 ha = 10^4 m^2
1 acre = 100 m^2
1 acre = 4047 m^2
1 acre = 43,560 ft^2
1 acre-ft = 43,560 ft^3
1 m^3 = 1000 L

Volume
1 ft^3 = 28.32 L
1 L = 0.03531 ft^3
1 L = 0.2642 gal
1 m^3 = 264.2 gal
1 ft^3 = 7.481 gal
1 m^3 = 35.31 ft^3

Velocity
1 m/s = 3.281 ft/s
1 mph = 1.467 ft/s
1 mph = 0.8684 knot
1 knot = 1.688 ft/s
1 km/h = 0.2778 m/s
1 km/h = 0.6214 mph

Force
1 lb = 4.448 x 10^5 dyne
1 lb = 32.17 pdl
1 lb = 0.4536 kg
1 N = 10^5 dyne
1 N = 0.2248 lb
1 kip = 1000 lb

Mass
1 oz = 28.35 g
1 lb = 0.4536 kg
1 kg = 2.205 lb
1 slug = 14.59 kg
1 slug = 32.17 lb

Work and Heat
1 BTU = 778.2 ft-lb
1 BTU = 1055 J
1 Cal = 3.088 ft-lb
1 J = 10^7 ergs
1 kJ = 0.9478 ft-lb
1 BTU = 0.2929 W·hr
1 ton = 12,000 BTU/hr
1 kWh = 3414 BTU
1 quad = 10^{15} BTU
1 therm = 10^5 BTU

Power
1 Hp = 550 ft-lb/s
1 HP = 33,000 ft-lb/min
1 Hp = 0.7067 BTU/s
1 Hp = 2545 BTU/hr
1 Hp = 745.7 W
1 W = 3.414 BTU/hr
1 kW = 1.341 Hp

Volume Flow Rate
1 cfm = 7.481 gal/min
1 cfm = 0.4719 L/s
1 m^3/s = 35.31 ft^3/s
1 m^3/s = 2119 cfm
1 gal/min = 0.1337 cfm

Torque
1 N·m = 10^7 dyne·cm
1 N·m = 0.7376 lb-ft
1 N·m = 10 197 g·cm
1 lb-ft = 1.356 N·m

Viscosity
1 lb-s/ft^2 = 478 poise
1 poise = 1 g/cm·s
1 N·s/m^2 = 0.02089 lb-s/ft^2

Pressure
1 atm = 14.7 psi
1 atm = 29.92 in Hg
1 atm = 33.93 ft H$_2$O
1 atm = 1.013 bar
1 atm = 1.033 kg/cm^2
1 atm = 101.3 kPa
1 psi = 2.036 in Hg
1 psi = 6.895 kPa
1 psi = 68 950 dyne/cm^2
1 ft H$_2$O = 0.4331 psi
1 kPa = 0.145 psi

PASS YOUR EXAM WITH THE ULTRA PE PREP SYSTEM

$25 OFF! YOUR PERSONAL SYSTEM!

Great Lakes Press 5-Step System is the best way to ensure exam readiness!

To create your own "Ultra PE Prep System," simply order a concise review, a past sample exam, a discounted handbook, and be sure you own and know how to use a powerful calculator—and you have the resources you need to pass the PE exam!

Step #1: Get the Best PE Review!

- ☐ *Principles & Practice of Civil / Mechanical / Electrical Engineering* edited by Merle C. Potter, PhD, PE, and written by teams of 9–10 veteran professors. These highly effective PE reviews are from the publisher of the best–selling FE/EIT review. Start with a great review…these are the only titles that concisely cover all critical aspects of the PE exam. Hundreds of excellent exam-simulating practice problems. 600-700 pp. (Solutions manuals to problems available separately.)

Step #2: Select a Past Sample Exam!

- ☐ *Official NCEES Sample Problems & Solutions.* These include actual past PE exam problems, all fully solved by official scorers. All PE exams are available in the latest printing. After your initial prep, take your sample exam (open book) to verify readiness or identify weaknesses. Use your review, handbook, and calculator to assist you—be sure you have easy access to all necessary information.

Step #3: Order a Handbook!

The PE exam is open–book, so be sure you have a current, all–inclusive handbook for easy, one–stop access to all necessary information.

- ☐ *Standard Handbook for Civil Engineers* by F. Merritt (4th ed., McGraw–Hill). New edition of the most thorough compilation of facts and figures. Best guide, best buy, and best seller since 1968!! 1,456 pp. List price: $150, GLP price: $119!
- ☐ *Mark's Standard Handbook for Mechanical Engineers* by E. Avalline (10th ed., McGraw–Hill). Practical advice and quick answers on all ME standards and practices. 1,792 pp. List price: $150, GLP price: $119!
- ☐ *Standard Handbook for Electrical Engineers* by Fink & Beaty (13th ed., McGraw–Hill). The premier EE reference. Up–to–date info in all areas of EE. 2,000 pp. List price: $125, GLP price: only $99!
- ☐ *Perry's Chemical Engineer's Handbook* by R. Perry (7th ed., McGraw–Hill). The best info on all aspects of ChemE…all in one place! The standard for over 40 years. 2,640 pp. List price: $150, GLP price: $119!
- ☐ *Structural Engineering Handbook* by Gaylord (4th ed., McGraw–Hill). Up–to–date information on all areas of Structural Engineering. Approx. 2,000 pp. List price: $125, GLP price: only $99!

Step #4: Don't Forget a Test–Beating Calculator!

- ☐ *HP 48GX Programmable Calculator.* Most states allow sophisticated calculators to be used when taking PE & FE exams. The HP 48 has hundreds of built–in equations you will need. GX plug–in cards offers exact FE/PE equations! Also, be sure you own *JumpStart the HP 48* (3rd ed.), a handy user's guide written to help engineers get the most out of these Hewlett Packard calculators.

Optional: ASCE Video Review Package (for PE Civil only)

- ☐ Newly updated with 23 hours of review from actual ASCE PE course. Includes comprehensive workbook. GLP throws in a free copy of our PE Civil Review and Solutions Manual—$93 value!

Step #5: Fill Out the Quick'n'Easy Order Form Below

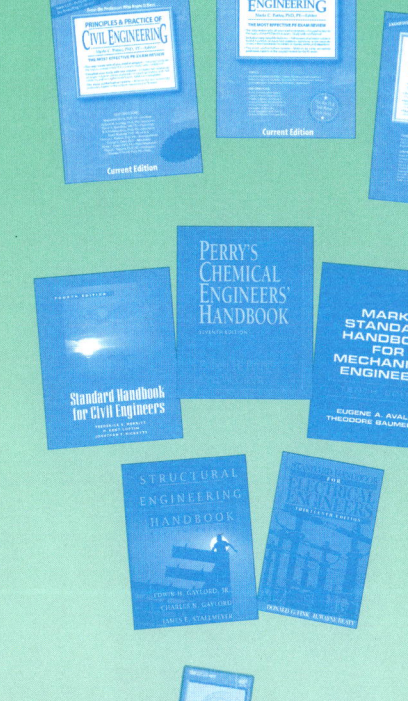

(please print)

Name_____

Address_____

City / State / ZIP _____

Phone with Area Code _____

Credit Card: ☐ [MC] ☐ [VISA] Exp. Date _____

Credit Card # _____

30–day Money Back Guarantee on GLP titles only.

Call **800-837-0201** or fax completed form to **636-273-6086**.

Or mail this form with credit card info or check or money order to:
Great Lakes Press
PO Box 550
Wildwood, MO 63040-0550

www.glpbooks.com   custserv@glpbooks.com   *Prices are subject to change.*

Description	Price	Ship	Quan.	Discount	Subtotal
PE CE / ME Review	$79.95 CE / $74.95 ME	$5		Package Savings on the Ultra PE Prep System! Take $25 off if ordering from Steps 1, 2, & 3!	
PE EE Review	$69.95	$5			
PE Solutions CE / ME / EE	$19.95	$3			
NCEES Sample PE Exam CE / ME / EE / ChemE / Enviro	$30	$5			
NCEES PE Exam, Structural	$45	$5			
CE / ME / EE / ChemE / Struct Handbooks	$99/$119	$7			
Calculator HP 48GX	$139.95	$7		N/A	
GX Card for FE / PE	$124.95	$7		N/A	
JumpStart the HP 48G/GX	$21.95	$3		N/A	
PE Civil Video Package	$595	$7		N/A	
Tax (MI & MO residents add 6.0% sales tax)					
TOTAL					

5/2001

CATALOG OF TITLES BY GREAT LAKES PRESS

FREE! Interactive CD With FE Titles!

Editor and Principal Author: Merle C. Potter, Ph.D., P.E.

Fundamentals of Engineering Review/Gen'l
10th edition, $54.95
The most effective review for the FE/EIT exam. Written by nine PhD professors for the A.M. and General P.M. sessions. This is the only review available that contains only what you need to pass—no more, no less! Includes full reviews of all 12 major general subjects; over 1,000 problems and full solutions, two fully–solved practice exams, and NCEES equation summaries. 669 pages. Includes coupon for FREE full-feature CD exams and StudyDirector™!

FE/EIT Discipline Reviews, 4 vols., 5th ed, $29.95
Covering all disciplines: CE, ME, EE, ChemE/IE, in four volumes. Each review teaches all essential equations in the *NCEES Reference Handbook*. Each vol. includes review, practice problems, solutions, & practice exam. Written by PhD teaching professors. 140-160 pp. each. Includes coupon for FREE exams CD with StudyDirector™!

NCEES FE Sample Exam & FE Supplied Reference Handbook, Exam $21.95, Handbook $10
Official 141-page sample exam from 2000. The *Reference Handbook* is a must for preparation—it is your only exam site resource and familiarity with it is essential. GLP reviews are keyed to the *Handbook*!

Jump Start the HP 48G/GX, 3rd edition, $21.95
Unique user manual written expressly for engineers. Superior to HP manual. Get the most out of your calculator! Makes passing tests much easier! Covers: Using Equations, Solved Problems, Special Applications, Writing Programs, Basic Statistics, File Transfer, Matrices, and more. 240 pages.

GRE, GMAT & GRE/Engineering Time•Savers™, $19.95 each
The most authoritative guides to these tests—authors are PhD teaching professors from major universities. Study same subjects as in other reviews…in much less time!
3 practice tests with solutions.

FULL REVIEWS FOR THE PRINCIPLES & PRACTICE OF ENGINEERING (PE) EXAMS:
Great Lakes Press' PE Exam reviews are written by 9-10 professors, experts in their fields. Reviews all major areas tested. Hundreds of excellent exam-simulating practice problems.

Principles & Practice of Civil Engineering Review, 4th edition, $79.95
The most concise coverage of all major subject areas tested. 650 pages. (Practice problems solutions manual, $19.95, 160 pp.)

Principles & Practice of Mechanical Engineering Review, 3rd ed., $74.95
The most concise coverage of all major subject areas tested. 586 pages. (Practice problems solutions manual, $19.95, 130 pp.)

Principles & Practice of Electrical Engineering Review, $69.95
The most concise coverage of all major subject areas tested. 397 pages. (Practice problems solutions manual, $19.95, 100 pp.)

Quick'n'Easy Order Form   *(please print)*

Name_____

Address_____

City / State / ZIP_____

Phone with Area Code_____

Credit Card: ❏ MC ❏ VISA   Exp. Date_____

Credit Card #_____

30–day Money Back Guarantee on GLP titles only.
Call **800-837-0201** or fax completed form to **636-273-6086**.

Or mail this form with credit card info or check or money order to:
Great Lakes Press
PO Box 550
Wildwood, MO 63040-0550

www.glpbooks.com   custserv@glpbooks.com   *Prices are subject to change.*

Description	Price	Quantity	Subtotal
Fundamentals of Engineering/General	$54.95		
FE/EIT Discipline: CE / ME / EE / ChemE/IE	$29.95		
NCEES: FE Exam / FE Handbook	$21.95/$10		
PE Review: CE / ME	$79.95 CE / $74.95 ME		
PE Review: EE	$69.95		
Solutions: CE / ME / EE	$19.95		
JumpStart HP 48 G/GX	$21.95		
GRE / GMAT Time•Saver	$19.95		
GRE / Engineering Review	$19.95		
Tax (MI & MO residents add 6% sales tax)			
Shipping ($7 first book + $1 ea. add'l)			
❏ Send a Catalog to a Friend or Colleague	**TOTAL**		

Name/Company_____

Address_____

5/2001

GREAT LAKES PRESS

READER REMARKS & REWARDS SURVEY

Your suggestions help us to improve this review continuously. As a way of saying thanks, we'll send you a FREE FE/EIT exams CD ($29.95 value) when you fill out and return this card. (Please include $5 shipping/handling.)

ABOUT YOU

Name _____

Address _____

City / State / ZIP _____

Phone / E–mail _____

Field / Position _____

❏ "You can tell them I said so!"   ❏ "Hey, send me that free CD!"
($5 ship/hndl, incl. check or CC#)

COMMENTS

_____

_____

_____

_____

Any content we missed? _____

_____

QUICK SURVEY

1) Your review?   F/G F/CE F/ME F/EE F/IE F/ChE P/CE P/ME P/EE

2) Overall, this book is:
   A   Too sketchy
   B   Just about right
   C   Too much material

3) The problems in this book are:
   A   Too easy
   B   Just about right
   C   Too difficult

4) The solutions are:
   A   Too sketchy
   B   Just about right
   C   Too much explanation

5) Did you participate in a review course?
   A   YES, Course location/name _____
   B   NO

6) Did you use other material in your preparation?
   Please list _____

7) How long since your undergraduate college graduation:
   _____ years   _____ haven't yet

8) Rank factors in order of influence on your initial appraisal of this book (1 being most important):
   _____ Price   _____ Reputation
   _____ Written by professors   _____ Presentation of material
   _____ Depth / Amount of material
   _____ Other _____

REWARDS FOR ERRATA!   *(Attach separate sheet if desired.)*

Think you found a mistake? We happily offer up to $3 for each error that has not already been discovered, depending on relevance

Error _____   Proposed Correction _____

Page Number _____   Problem or Example Number _____

SEND INFO TO A FRIEND, COLLEAGUE, OR COMPANY MANAGER

The following people would appreciate a *one–time* mailing of a catalog of your FE & PE resources.

Name _____   Name _____   Name _____

Address _____   Address _____   Address _____

_____   _____   _____

City _____   City _____   City _____

State / ZIP _____   State / ZIP _____   State / ZIP _____

❏ Civil   ❏ Mechanical   ❏ Civil   ❏ Mechanical   ❏ Civil   ❏ Mechanical
❏ Electrical ❏ Other   ❏ Electrical ❏ Other   ❏ Electrical ❏ Other

TO RECEIVE *FREE* CD ...a $29.95 value!...fill out post-paid form, return with credit card info or check payable to "Great Lakes Press," for $5 shipping/handling.

Credit card # _____   Exp. Date: _____

CC Billing Address: _____

Cut along this line, then fold, tape and mail

5/2001

Hey! Send me my FREE FE Exams CD!

☐ ✓ Check this box, fill out entire form on reverse side...
...then tear, fold and send in this whole postpaid card!
(please allow 2 weeks for delivery)

• CD includes 6 solved exams, Study-Director™
...and much more!

Fold here second ↓

Just fill out, fold, tape & drop this survey card into any mailbox!

BUSINESS REPLY MAIL
FIRST-CLASS MAIL   PERMIT NO 71   GROVER, MO
POSTAGE WILL BE PAID BY ADDRESSEE

**GREAT LAKES PRESS
PO BOX 550
WILDWOOD, MO 63040-9913**

NO POSTAGE
NECESSARY
IF MAILED
IN THE
UNITED STATES

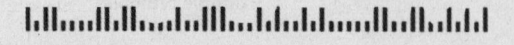

Fold here first ↑

Cut along this line, then fold, tape and mail